COSMIC SPACE IS GOD AND PHYSICAL UNIVERSE IS GOD'S DREAM - 2

DR. CHANDRA BHAN GUPTA

B.Sc. (Lko.), M.B.B.S. (Lko.),

M.D. THESIS (MED). (ALLD).,M.R.C.P.(UK),

F.R.C.P. (Edin.), F.R.C.P.(Glasg.),

E.C.F.M.G. CERTIFICATE (U.S.A.)

ISBN: 978 166 139 306 9

With Love to My Children

Nirupama

Sujata

Anil

TABLE OF CONTENTS

ACKNOWLEDGEMENT

Let me firstly express my feelings of gratitude to all those who have been instrumental in helping me create this book.

Secondly, I am extremely thankful to my wife Diana, for her help in the preparation of the book despite her various other responsibilities, and to my daughter Sujata for her help. Thirdly, I extend my grateful thanks to my son-in-law, Sunil, whose invaluable technical and practical assistance was proffered notwithstanding his extremely busy schedule.

I would also like to mention many Adwaitic thinkers, scholars, teachers, seekers and students, who constantly, urged me to write this book, after reading my previous books entitled

'Adwaita Rahasya : Secrets of Creation Revealed'
'Space is The Mind of God : A Scientific Explanation of God and His Abode'
'Cosmic Space Is God and Physical Universe is God's Dream',

Finally, I express my sense of intense gratitude to Cosmic Space, who fills the whole cosmos, for choosing me as His instrument to pen this book.

TRANSCENDENTAL CONSCIOUSNESS - 1

The status of 'Transcendental Consciousness' is an earned one; earned as a result of one's effort.

This highly prized appellation one gains after striving a great deal. It is not procured easily and without pain. On the contrary, during the process of its attainment, one has to undergo not only repeated failures and disappointments but also a lot of suffering. However, the end result of it all is a highly satisfying one because it propels the person concerned one step higher on the very difficult path of understanding God.

On the evolutionary ladder of man's spiritual and supra-spiritual development, the rank of 'Transcendental Consciousness' is only an intermediate step.

Following the attainment of the status of 'Transcendental Consciousness', mankind's consciousness becomes free of a very ancient delusion called the delusion of 'body consciousness.'

From time immemorial, delusion of 'body consciousness' has kept mankind imprisoned within the confines of an erroneous belief that the identity of a person comprises solely of a person's physical body and nothing else, the physical body which is, as its name implies, merely a physical thing and thus has no innate capacity to think, feel, perceive and be self-aware in the absence of its primeval and priceless companion called the consciousness.

The identity of one's 'true self" or, more precisely, the identity of one's 'I' comprises of one's consciousness only and not the body or, for that matter, anything else. This consciousness is intrinsically a thinking, feeling, perceiving and self-aware entity who merely lives in the physical body for a short while at the behest of god, nothing more nothing less.

Prior to its victory over the delusion of 'body consciousness', man's consciousness is addressed merely as an 'infra-transcendental consciousness'. It is only after it has conquered its delusion of 'body consciousness' and has entered into the state of 'consciousnessbal consciousness' that it becomes entitled to be called, 'transcendental consciousness'.

Attainment of 'consciousnessbal consciousness' thrusts man into the domain of the fundamental truth that he is not the 'physical object' called the body. Instead, he is the 'non-physical subject' called the consciousness.

TRANSCENDENTAL CONSCIOUSNESS - 2

A **'Transcendental Consciousness'** is an unusual human consciousness. Its unusualness lies in the fact that it has been able to achieve victory in a very arduous task called the task of overpowering the 'maya' or delusion of 'body consciousness.'

'Maya' or delusion of 'body consciousness' consists of a false belief on a person's part that its true identity consists of the physical body.

A **'Transcendental Consciousness'** is exceptional , not merely in the sense that it has been victorious in overcoming the 'maya' or delusion of 'body consciousness' but also in the sense that it has been triumphant in gaining a new insight into the true nature of its 'I' or into the true nature of its 'True Self ' that its 'I' or the 'True Self ' comprises of that inherently sentient or thinking, feeling, perceiving and self-aware entity called consciousness which addresses itself universally as 'I' and which is non-physical in nature, in contrast to body which is physical in nature and is inherently insentient or non-thinking, non-feeling, non-perceiving and non-self-aware entity in the absence of its in-dwelling **sentient** companion called consciousness.

Such a **'Transcendental Consciousness'** must not feel satiated at this lone achievement only, even though, without a doubt, this lone achievement itself is an extremely formidable accomplishment in its own right.

Instead, it must forge ahead still further by gaining triumph in some more assignments. These assignments consist of apprehending plus internalizing some more but much higher truths which are chronicled or set out below:-

A **Transcendental Consciousness** must appreciate and internalize the fact that despite being a sentient or thinking, feeling, perceiving and self-aware plus non-physical **consciousness** in the manner of god, creator, maker or progenitor of the current cosmos, it merely is a **'drop'** of **consciousness,** albeit an absolutely pure or pristine **'drop'** of **consciousness,** derived from the ubiquitous and infinite **ocean of consciousness** called **cosmic space** which, in turn, is nothing but the **expanded, distended, dilated** or **inflated** form or version of the **dimensionless** form or version of god's or creator's **consciousness** and this **dimensionless** form or version of god's or creator's **consciousness** was its **original** form or version which existed some 13.7 billion light years ago only and does not exist now. God's or creator's **consciousness** now exists instead, in its **expanded, distended, dilated** or **inflated** form or version as the ubiquitous and infinite, **field of consciousness** or **ocean of consciousness** called **cosmic space.** In other words, a **transcendental consciousness** must appreciate or apprehend plus digest, absorb and internalize the fact that the **consciousness** of god, creator, maker or progenitor of the current cosmos presently exists in the form of the current, non-physical **cosmic space.**

2. A **'Transcendental Consciousness'** must also appreciate or apprehend plus digest, absorb and internalize the fact that by being a sentient or thinking, feeling, perceiving, self-aware, non-physical plus an absolutely **pure** or **pristine 'drop'** of **consciousness,** derived from the ubiquitous and infinite **ocean of consciousness** called **cosmic space,** it innately and absolutely is **timeless, immortal** or **eternal** in the manner god, creator, maker or progenitor of the current cosmos is.

Additionally, the **Transcendental Concsiousness** must also grasp the truth that god, creator, maker or progenitor of the current cosmos is none other than the amazing, ubiquitous and infinite **ocean of consciousness** called **cosmic space.** Hence, **cosmic space** is not some insentient, unthinking, unfeeling, lacking perception, lacking sensation, lacking consciousness, inert and dead to the world, non-physical thing which happened to be there by serendipity in order to spatially accommodate the current **3-D** or **three-dimensional** physical cosmos inside itself. Instead, it is the amazing, sentient or thinking, feeling, perceiving and self-aware, non-physical **consciousness** of god or creator of the current cosmos.

In other words, a **Transcendental Consciousness** must comprehend plus digest, absorb and internalize the truth that by being a **pure** or **pristine,** non-physical **'drop'** of **consciousness,** derived from the amazing, ubiquitous and infinite **ocean of consciousness** called **cosmic space,** it too is **deathless, imperishable** or **indestructible** in the manner god, creator, maker or progenitor of the current cosmos is which is none other than this very same entity called **cosmic space.**

By way of explanation, a **Transcendental Consciousness** must appreciate or apprehend plus accept, absorb and assimilate or internalize the truth that the current **cosmic space** is the **mind of god** or the **mind of creator, maker** or **progenitor** of the current cosmos and this amazing, ubiquitous and infinite **mind of god** or **mind of creator, maker** or **progenitor** of the current cosmos presently is in its **expanded, distended, dilated or inflated form** or **version** in order to give rise to a **mind-space** or **consciousnessbal-space** inside itself, or in order to give rise to enough **room, area** or **acreage** inside its **mind** or **consciousness,** for the **spatial placement** and **existence** of its current **3-D** or **three-dimensional daydream** or **reverie** called the **physical cosmos.**

A **Transcendental Consciousness** must also take cognizance of the fact that the physical bodies of the current human consciousnesses themselves are a constituent part of this 3-D or three-dimensional daydream of god called physical cosmos. On account of being a constituent part of this daydream of god, creator, maker or progenitor of the current cosmos, human beings have an inbuilt blind spot inside their mind, consciousness or awareness which makes them ignorant of the true nature of the current cosmos and prompts them to label the current cosmos as one hundred percent real and physical cosmos.

But the truth is that the current **physical cosmos** is nothing of the sort. Instead, it is a pure and simple **daydream** of god and nothing else. And therefore, it is composed of **daydream-stuff** of god or, if is preferred, it is composed **mind-stuff** of god, creator, maker or progenitor of the current cosmos. God, creator, maker or progenitor of the current cosmos presently exists as the amazing, ubiquitous and the infinite **ocean of consciousness** called **cosmic space** which, in turn, is nothing but the **expanded, distended, dilated** or **inflated** form or version of god's original **dimensionless consciousness** which existed some 13.7 billion light years ago only and not now because, as said just above, it now exists in its **expanded, distended, dilated** or **inflated** form or version called **cosmic space.**

What has been said above can be put in another way.

A **Transcendental Consciousness** is a **human consciousness** which has apprehended, accepted, digested, absorbed and assimilated or incorporated within **its being** the following truths.

God, creator, maker or progenitor of the current cosmos existed as an incredible or awe-inspiring plus **the only one**

of its kind or unique, **timeless, bodiless** and **dimensionless consciousness** of infinite intelligence, imagination and emotion some 13.7 billion light years ago.

This **timeless, bodiless** and **dimensionless** form or version of **consciousness** of god, creator, maker or progenitor of the current cosmos was its **original** form or version. That is to say, the **consciousness** of god, creator, maker or progenitor of the current cosmos existed in its **original** form or version right back, approximately 13.7 billion light years ago only and not anytime recently.

The **original** form or version of **consciousness** of god, creator, maker or progenitor of the current cosmos is chronicled or recounted as its **dimensionless** form or version.

Therefore, the whole description of the **original** form or version of god, creator, maker or progenitor of the current cosmos is set out or recounted as the **dimensionless consciousness** of god or creator or, more to the point, as the **bodiless** and **dimensionless consciousness** of god or creator or, better still, as the incredible or awe-inspiring and **the only one of its kind** or unique, **timeless, bodiless** and **dimensionless consciousness** of god, creator, maker or progenitor of the current cosmos.

God, creator, maker or progenitor of the current cosmos presently exists, not in its **original** form or version, that is to say, not in its **dimensionless** form or version or, better still, not in the form or version of **dimensionless consciousness.** Instead, it presently exists as the amazing, ubiquitous, and infinite, **3-D** or **three-dimensional, field of consciousness** or **ocean of consciousness,** called **cosmic space.**

The **consciousness** of god, creator, maker, or progenitor of the current cosmos has **metamorphosed** itself into its present form or version called the amazing, ubiquitous, and infinite, **3-D** or **three-dimensional, field of consciousness** or **ocean of consciousness** aka **cosmic space,** through the process of **expansion, distention, dilation** or **inflation** of its **original** form or version or, better still, of its **dimensionless** form or version so that it can have enough room or space inside **itself** or, better still, so that it can have enough room or space inside its **dimensionless-self** or, if it is preferred, so that it can have enough room or space inside its **dimensionless mind, consciousness** or **awareness,** in order to **spatially** accommodate its subsequently produced, **3-D** or **three-dimensional daydream, reverie, imagery, dreamry, fantasy** or **phantosmagoria,** whom the current **human consciousnesses** ignorantly, nonsensically, unintelligently or without care label as one hundred percent real, **physical cosmos.**

The **physical bodies** of the present-day **human consciousnesses** themselves are a **constituent part** of god's or creator's current **daydream, reverie, imagery, dreamry, fantasy** or **phantosmagoria** which is ignorantly or without care labeled as one hundred percent real, **physical-cosmos** by the current **human consciousnesses.**

Thus, the **'mind of god'** or the **'mind of creator'** is none other than the entity which is ignorantly, nonsensically, unintelligently or without care called **cosmic space** by the present-day **human consciousnesses.**

Inside this **'mind of god'** or **'mind of creator'** aka the present-day **cosmic space,** the current **physical cosmos** is floating, wafting or levitating plus whirling, twirling or spiraling as a mere **daydream, reverie, imagery, dreamry, fantasy** or **phantasmagoria** of this **'mind of god'** or **'mind of creator '.**

Therefore, the current physical cosmos is not real and physical in the sense the current human consciousnesses think and feel it is real and physical. Instead, it is real and physical in the sense all dream sleep state cosmoses seem real and physical to the selfsame or the very same or one and the same human consciousnesses when they are in their dream sleep state.

The present-day human consciousnesses think and feel that the current physical cosmos is absolutely real and physical because they themselves are a constituent part of this daydream, reverie, imagery, dreamry, fantasy or phantosmagoria of god or creator. As a result, they do have an independent yardstick to judge what is absolutely real and what is only relatively real.

THE SUPRA - TRANSCENDENTAL CONSCIOUSNESS

The appellation supra-transcendental consciousness is another epithet or moniker of god or the creator of the current physical cosmos.

This supra-transcendental consciousness, god or the creator of the current physical cosmos is timeless. It is the only one of its kind, constant, eternal, ceaseless, deathless, incessant and immortal awareness or consciousness of infinite intelligence, imagination and emotion. It is formless, dimensionless, non-physical, non-material or incorporeal. That is to say, it is without physical or material body, form, or substance and therefore, is formless and dimensionless. This can be put in another way. The supra-transcendental consciousness, god, or the creator of the current physical cosmos is a unique, timeless, dimensionless and bodiless or disembodied, disbodied, or discarnate awareness or consciousness of infinite intelligence, imagination and emotion.

There is nothing in the present-day physical cosmos which is timeless, constant, eternal, ceaseless, deathless, incessant, or immortal plus formless, dimensionless and bodiless or disembodied, disbodied, or discarnate in the manner, supra-transcendental consciousness, god or the creator of the current physical cosmos is. All the objects of the present-day physical cosmos, without exception, are such things which last only for a short while. In other words, they universally are evanescent, ephemeral, passing, fleeting, transient, temporary, unstable, changeable, mutable, momentary, mortal, short-lived, or here today and gone tomorrow kind of things. Additionally, they all, everywhere are in concretized, materialized, manifested, material, corporeal, incarnate, tangible, embodied, personified, or bodily forms. They all, therefore, are dimensional things, meaning thereby, they all, therefore, are in compulsory, mandatory, predetermined, preordained, inevitable or inescapable need of a pre-existent space or, more to the point, a pre-existent cosmic space for their birth and subsequent spatial placement and existent.

What has been said above can be put in another way.

In the absence of the pre-existent space or, more to the point, the pre-existent cosmic space, the concretized, materialized, manifested, embodied, incarnate, bodily, physical, material, or corporeal component, constituent or ingredient of the current cosmos namely, the physical matter/physical energy duo will never take birth. In other words, the current space or, more to the point, the current cosmic space is the sine qua non or the indispensable pre-condition or pre-requisite for the birth and subsequent existence or continuance of the concretized, materialized, manifested, embodied, incarnate, bodily, physical, material, or corporeal cosmos namely the current physical matter/physical energy duo.

Let one explain further.

The current, concretized, materialized, manifested, embodied, incarnate, bodily, physical, material, or corporeal cosmos i.e. the current physical matter/physical energy duo, is a created thing or, to be absolutely forthright, frank, or candid, is a daydream-stuff composed thing, brought into being by its creator aka god or the supra-transcendental consciousness. The birth of a created thing or, to be absolutely forthright, frank, or candid, of a daydream-stuff composed thing such as the current physical matter/physical energy duo, cannot occur without the wish or will of its creator or, better still, without the wish or will of its daydreamer namely, god, or the supra-transcendental consciousness. And the creator, daydreamer or god of the current, concretized, materialized, manifested, embodied, incarnate, bodily, physical, material, or corporeal cosmos i.e. the current physical matter/physical energy duo is none other than the current space or, more to the point , the current cosmic space which in fact, is nothing but god, creator, maker or daydreamer of the current, concretized, materialized, manifested, embodied, incarnate, bodily, physical, material or corporeal cosmos aka the current physical matter/physical energy duo.

The concretized, materialized, manifested, embodied, incarnate, bodily, physical, material or corporeal component or constituent of the current cosmos i.e. the physical matter/physical energy duo, is only one amongst the overall or all in all, three ingredients, components or constituents which, in concert, cooperation or collaboration are responsible for the construction or creation of the current cosmos. The other two ingredients, constituents or components, which in concert, cooperation or collaboration with the concretized, materialized, manifested, embodied, incarnate, bodily, physical, material or corporeal constituent or component i.e. physical matter/physical energy duo, have given rise to the current cosmos, are cosmic space on one hand and human plus other embodied consciousnesses of the current cosmos on the other.

Out of the three constituents or components, which conjointly or collectively compose the current cosmos, only the corporeal or the concretized constituent or component of the current cosmos which is constantly changing, changeable, movable, mutable, unsteady, unstable, fleeting, floating, evanescent, ephemeral, transient, temporary, momentary, mortal, passing, short-lived, chameleon-like, or here today and gone tomorrow type.

What has been said above can be put in another way.

Out of the three constituents or components, which conjointly or collectively compose the current cosmos, only the materialized, manifested, embodied, incarnate, bodily, physical, or material constituent or component of the current cosmos is evanescent, ephemeral, fleeting, floating, transient, temporary, unstable, changeable, mutable, momentary, mortal, short-lived or here today and gone tomorrow type.

The other two constituents or components of the current cosmos namely the embodied consciousnesses, (for example, human consciousnesses) on one hand and the cosmic space on the other are both timeless, immortal or eternal. These two constituents or components of the current cosmos are not time-bound, transient, or temporary in the manner of the physical matter/physical energy duo dyad, doublet or twosome i.e. in the manner of the materialized, manifested, embodied, incarnate, bodily, physical, or material constituent or component of the current cosmos.

The current cosmic space is timeless, immortal or eternal due to the fact that it is nothing but the expanded, distended, dilated or inflated form or version of the dimensionless form or version or, the original form or version of the timeless, immortal or eternal god, creator or maker of the current cosmos.

All the embodied consciousnesses of the current cosmos, for example, all the human consciousnesses of the current cosmos are timeless, immortal or eternal because they are absolutely unmodified, unaltered, unvitiated or, absolutely pristine or pure and uncorrupted form or version of a section, segment, part, or portion of the ubiquitous and infinite ocean of consciousness aka cosmic space which is nothing but the expanded, distended, dilated or inflated form or version of the dimensionless form or version, that is to say, of the original form or version of the timeless, immortal or eternal god, creator or maker of the current cosmos.

After the inevitable death of their respective, concretized, materialized, manifested, material, corporeal, physical or incarnate bodies, all the human consciousnesses or, better still, all the previously embodied consciousnesses, including the human consciousnesses of the current cosmos, immediately unite back with or merge back into the ubiquitous and infinite ocean of consciousness called cosmic space which is nothing but the expanded, distended, dilated or inflated form or version of the dimensionless form or version, that is to say, of the original form or version of the timeless, immortal or eternal god, creator or maker of the current cosmos.

Here, one must reassure humanity with all the certainty, certitude, conviction or confidence one has at one's command that it is only their physical bodies which are mortal or subject to death (and therefore will die one day) and not their consciousnesses. Their consciousnesses are as timeless, immortal or eternal as the current cosmic space is and the current cosmic space is nothing but the expanded, distended, dilated or inflated form or version of the dimensionless form or version, that is to say, of the original form or version of the timeless, immortal or eternal god, creator, maker, or progenitor of the current cosmos.

Here it will not be amiss, or wrong to remind humanity once again that the current cosmic space is nothing but the expanded, distended, dilated, or inflated form or version of the dimensionless form or version or, the original form or version of the timeless, immortal or eternal god, creator, maker, or progenitor of the current cosmos.

In contrast to all the embodied consciousnesses of the current cosmos, (for example, all the embodied human consciousness of the current cosmos) ; all of whom are timeless, immortal or eternal in the manner of the timeless, immortal or eternal god, creator, or maker of the current cosmos which is none other than the entity called the current cosmic space ; the physical bodies of all the embodied consciousnesses of the current cosmos, (for example, the physical bodies of all the embodied human consciousnesses of the current cosmos), are time-bound, mortal or subject to death. This is so because these physical bodied have been constructed, created, or put together out a condensed, compressed or compacted or, if it is preferred, out of a congealed, or consolidated, or, better still, out of a concretized or solidified section, segment, part or portion of the ubiquitous and infinite ocean of consciousness aka cosmic space and not out of an absolutely unmodified, unaltered, unvitiated or, absolutely pristine, or pure and uncorrupted form or version of a section, segment, part, or portion of the ubiquitous and infinite ocean of consciousness aka cosmic space, as is the case with regards to all the embodied consciousnesses of the current cosmos.

Thus, all the embodied consciousnesses of the current cosmos, including all the embodied human consciousnesses of the current cosmos, are timeless, immortal or eternal in the manner, the only one of its kind, timeless, dimensionless, bodiless or non-embodied, disembodied, or discarnate consciousness of god, creator, maker or progenitor of the current cosmos is.

THE SUPRA-TRANSCENDENTAL CONSCIOUSNESS AND THE DREAM SLEEP STATE OF HUMAN CONSCIOUSNESSES

The supra-transcendental consciousness, that is to say, the consciousness of god, creator, maker, or progenitor of the current physical cosmos is the only one of its kind, or unique, timeless, eternal or immortal, non-embodied, disembodied, disbodied, discarnate, or bodiless consciousness.

The consciousnesses of human beings on the other hand, are timeless, immortal, or eternal, embodied, personified, corporealized, materialised, manifested, or incarnate consciousnesses.

If the embodied human consciousnesses of the current physical cosmos, directly, personally, empirically, factually or at first-hand, wish or desire to observe, see, know, and experience, the only one of its kind or unique, timeless, immortal, or eternal, non-embodied, disembodied, disbodied, discarnate, or bodiless, supra-transcendental consciousness or the consciousness of god, creator, maker, or progenitor of the current physical cosmos in their fully awake or wide awake state, then they have to look no further than at the incredible or awe-inspiring plus the ubiquitous and infinite, **3-D** or **three-dimensional, field of consciousness** or **ocean of consciousness** called **cosmic space** which is nothing but the 3-D or three-dimensional form or version of the **dimensionless form** or **version,** or the **original form** or **version** of the **supra-transcendental consciousness** or the consciousness of god, creator, maker or progenitor of the current physical cosmos.

In their fully awake or wide -awake state, the embodied human consciousnesses are fully aware or conscious, all around themselves, of the existence or presence of their circumscribed, delineated, or limited, 3-D or three-dimensional physical or material body.

Therefore, when the embodied human consciousnesses are in their fully awake or wide awake state, and observe or see, with the aid of their physical eyes, the non-physical, non-material, unphysical, or immaterial, ubiquitous and infinite, 3-D or three-dimensional cosmic space aka supra-transcendental consciousness aka god aka creator, maker or progenitor of the current physical cosmos, they, at the very same moment, also observe, see or perceive, and, are aware or conscious , all around themselves, of the existence or presence of their physical or material, circumscribed, delineated, or limited, 3-D or three-dimensional body.

However, the embodied human consciousnesses, willy-nilly, unavoidably, or perforce, are also given, albeit only for a 'brief period' every night, during their dream sleep state, a subjective-experience of bodiless-ness, non-embodied-ness, disembodied-ness, disbodied-ness, or, discarnate-ness by their 'source' or 'fountainhead' namely, the supra-transcendental consciousness aka **cosmic space** aka god, creator, maker or progenitor of the current physical cosmos. But this 'brief period' (for which the embodied human consciousnesses, during their dream sleep state, are given every night, a subjective-experience of bodiless-ness, non-embodied-ness, disembodied-ness, disbodied-ness, or discarnate-ness, by their 'source' or 'fountainhead' aka supra-transcendental consciousness aka **cosmic space** aka god, creator, maker or progenitor of the current physical cosmos) is a 'brief period', only from the standpoint of the timescale of their **fully-awake** or **wide-awake-state,** and not from the standpoint of the timescale of their **dream-sleep-state.**

From the standpoint of the timescale of the **dream-sleep-state,** the above mentioned, 'brief period' of the **subjective-experience** of non-embodied-ness, disembodied-ness, disbodied-ness, discarnate-ness or bodiless-ness, imparted to human consciousnesses during their **dream-sleep-state,** by their 'source' or 'fountainhead' namely, the **cosmic space** aka supra-transcendental consciousness aka god, creator, maker or progenitor of the current cosmos, is as prolonged or as long as the **existence-span** of their **dream-sleep-state,** which, at the time, seems to the human consciousnesses of the **dream-sleep-state,** as prolonged or as long as the **existence-span** of their **fully-awake** or **wide-awake-state.** This illustrates the truth that the **subjective-feeling** called **'time',** experienced by human consciousnesses is always **'relative'** or **'comparative'** and never **'definitive'** or **'absolutive'** (i.e. **absolute).**

Thus, through the experience of their dream sleep state every night, each and every embodied human consciousness of the **fully-awake** or **wide-awake-state,** is able to taste, sample, or savour, directly, personally, empirically, factually or at first-hand, the **cosmic-space-like** aka **god-like** state of non-embodied-ness, disembodied-ness, disbodied-ness, discarnate-ness or bodiless-ness, which is not available to it in its **fully-awake** or **wide-awake-state.** In other words, in its dream-sleep-state each and every human consciousness becomes a **mini-replica** of current **cosmic space** aka **god** because in this state, each and every human consciousness becomes subjectively **non-embodied, dis-embodied, disbodied, discarnate** or **bodiless** as well as **3-D** or **three-dimensional** in configuration in the manner current **cosmic space** aka **god, creator, maker** or progenitor of the current **physical cosmos** is.

What has been said above can be put in another way.

Through its nightly experience of dream sleep state, each and every human consciousness, when it returns to its wakeful state, should be able to bring back to memory the experience of its dream-sleep-state and through it, should be able to taste, sample, or savour, directly, personally, empirically, factually or at first-hand within its consciousness, the state of non-embodied-ness, disembodied-ness, disbodied-ness, discarnate-ness or bodiless-ness of **cosmic space** aka supra-transcendental consciousness aka god, creator, maker or progenitor of the current physical cosmos. The non-embodied-ness, disembodied-ness, disbodied-ness, discarnate-ness or bodiless-ness is one of the innate attributes, traits, or qualities of **cosmic space** aka supra-transcendental consciousness aka god, creator, maker or progenitor of the current physical cosmos. The other innate attribute, trait or quality of **cosmic space** aka supra-transcendental consciousness aka god, creator, maker or progenitor of the current physical cosmos is **timelessness** or **immortality.**

However, one thing must be pointed out here to human consciousnesses. And the point is that they cannot directly, personally, empirically, factually or at first-hand experience their own innate, **cosmic-space-like** or **god-like,** timelessness or immortality at the present moment. That is to say, human consciousnesses cannot directly, personally, empirically, factually or at first-hand experience their own innate, **cosmic-space-like** or **god-like** timelessness or immortality, so long as they exist in their current, physically-embodied form or version as one of the **constituent-part** of the present-day, **3-D** or **three-dimensional,** time-bound or mortal **physical cosmos** because this experience of **cosmic-space-like** or **god-like** timelessness or immortality does not exist anywhere in the current, innately transient, temporary, or time-bound, **3-D** or **three-dimensional, physical cosmos.**

Apart from the non-physical, non-material, unphysical, or immaterial, ubiquitous, and infinite, 3-D or three-dimensional field of consciousness or ocean of consciousness aka cosmic space aka god, creator, maker or progenitor of the current physical cosmos, the non-physical, non-material, unphysical, or immaterial, physically-embodied consciousnesses of the current physical cosmos, for example, human consciousnesses of the current physical cosmos, are the only other entities in the current physical cosmos which are innately timeless, immortal, or, eternal.

The innate, timelessness or immortality of **non-physical, non-material, unphysical, or immaterial,** physically-embodied consciousnesses of the current physical cosmos, for example, human consciousnesses of the current physical cosmos, will only be actualised or realised, following the **disembodiment** or **discarnation** of these **non-physical, non-material, unphysical, or immaterial** consciousnesses of the current physical cosmos, for example, human consciousnesses of the current physical cosmos, which will take place only at the moment of death of their transient, temporary, or, time-bound, physical, or material-bodies, or, if it is preferred, objective-bodies, at which moment, they, namely, the **disembodied** or **discarnate, non-physical, non-material, unphysical, or**

immaterial consciousnesses of the current physical cosmos, for example, human consciousnesses of the current physical cosmos, will merge-back or meld-back into their 'source', or 'fountainhead', namely, the present-day, **non-physical, non-material, unphysical, or immaterial,** ubiquitous, & infinite, **3-D** or **three-dimensional field of consciousness** or **ocean of consciousness** aka **cosmic space** aka **god,** creator, maker, or progenitor of the current physical cosmos.

What has been said above can be put in another way.

The innate, timelessness or immortality of **non-physical, non-material, unphysical, or immaterial,** physically-embodied consciousnesses of the current physical cosmos, for example, human consciousnesses of the current physical cosmos, will only be actualised or realised, following their **disembodiment** or **discarnation,** which will take place only at the moment of death of their transient, temporary, or, time-bound, physical, or material-bodies, or, if it is preferred, objective-bodies.

Following the **death** of the transient, temporary, or, time-bound, physical, or material-bodies of the **non-physical, non-material, unphysical, or immaterial,** consciousnesses of the current physical cosmos, for example, human consciousnesses of the current physical cosmos, or, if it is preferred, following the **disembodiment** or **discarnation** of the **non-physical, non-material, unphysical, or immaterial,** consciousnesses of the current physical cosmos, for example human consciousnesses of the current physical cosmos, the latter, namely, the timeless or immortal, **non-physical, non-material, unphysical, or immaterial,** consciousnesses of the current physical cosmos, for example human consciousnesses of the current physical cosmos, will merge-back or meld-back into their 'source', or 'fountainhead' which is none other than the present-day, **non-physical, non-material, unphysical, or immaterial,** ubiquitous, and infinite, **3-D** or **three-dimensional field of consciousness** or **ocean of consciousness** aka **cosmic space** aka **god,** creator, maker, or progenitor of the current physical cosmos.

The verity, reality, or, certainly of innate-immortality, or timelessness of non-physical, non-material, unphysical, or immaterial, ubiquitous, and infinite, 3-D or three-dimensional field of consciousness or ocean of consciousness aka cosmic space aka supra-transcendental consciousness aka god, creator, maker, or progenitor of the current physical cosmos on one hand, and that of all the embodied or incarnate consciousnesses of the current physical cosmos, for example, all the human consciousnesses of the current physical cosmos on the other, will have to be discovered by the current embodied or incarnate human consciousness of the current physical cosmos themselves, through the dint of their own effort, via the medium of consciousnessbal scrutiny, analysis, assessment or evaluation. There is no short-cut available to them in this regard.

In their dream sleep state, all embodied human consciousnesses are an aware being in exactly the same manner as they are in their fully awake or wide- awake state. But in this state, in stark contrast to their fully awake or wide- awake state, they are not aware of their physical body of which they are aware in their fully awake or wide-awake state nor are they aware of the **3-D** or **three-dimensional** physical cosmos of which they are aware when they are in their fully awake or wide -awake state.

Thus, the dream-sleep-state of human consciousnesses is an extraordinary state from two points of view which are elaborated below.

Firstly, the feeling or the perception of physical-embodied-ness, from which all human consciousnesses, inescapably, or perforce, suffer, during their fully-awake, or wide-awake-state, or, with which all human consciousnesses, inescapably, or perforce, are saddled with or encumbered with during their fully-awake or wide-awake-state, is temporarily removed, withdrawn, or, taken away from them during their dream-sleep state by their 'source' or 'fountainhead' namely, the **non-physical, non-material, unphysical, or immaterial,** ubiquitous, and infinite, **3-D** or **three-dimensional field of consciousness** or **ocean of consciousness** aka **cosmic space** aka **supra-transcendental consciousness** aka **god,** creator, maker, or progenitor of the current physical cosmos.

As a result, during the brief period of their dream-sleep-state, all physically-embodied human-consciousnesses of the fully-awake or wide-awake-state, become subjectively, non-embodied, disembodied, disbodied, discarnate, or bodiless in the manner their 'source' or 'fountainhead' namely the **non-physical, non-material, unphysical, or immaterial,** ubiquitous, and infinite, **3-D** or **three-**

dimensional field of consciousness or **ocean of consciousness** aka **cosmic space** aka **supra-transcendental consciousness** aka **god,** creator, maker, or progenitor of the current physical cosmos **eternally** or **perpetually** is.

Secondly, such subjectively, non-embodied, disembodied, disbodied, discarnate, or bodiless, human consciousnesses of the dream-sleep-state, directly, or, at first hand, observe or see, or, if it is preferred, directly, or, at first hand, perceive and experience a non-physical or non-material, **3-D** or **three-dimensional, dreamal-cosmic-space** which is situated or seated inside their very own subjectively, non-embodied, disembodied, disbodied, discarnate, or bodiless consciousnesses of their dream-sleep-state.Additionally, the subjectively, non-embodied, disembodied, disbodied, discarnate, or bodiless human consciousnesses of the dream-sleep-state, directly or, at first hand, also observe, see, perceive and experience, during their dream-sleep-sleep, a **physical-looking, 3-D** or **three-dimensional, dreamal-cosmos** or **subjective-cosmos,** which is dismissed, discarded, scorned, snubbed, scoffed at, sneered at, brushed off, brushed aside, discounted or rebuffed as a mere **dream** of mankind's **dream-sleep-state** and therefore of no importance, significance or value, by the human consciousnesses of the fully-awake or wide-awake-state.

This **physical-looking, 3-D** or **three-dimensional, dreamal-cosmos** or **subjective-cosmos** (which is dismissed, discarded, scorned, snubbed, scoffed at, sneered at, brushed off, brushed aside, discounted or rebuffed as a mere **dream** of mankind's **dream-sleep-state** and therefore of no importance, significance or value, by the human consciousnesses of the fully-awake or wide-awake-state), is spatially or territorially seated or situated inside a non-physical, non-material, unphysical, or immaterial, **3-D** or **three-dimensional, dreamal-cosmic-space** which, in turn, is itself seated or situated inside the subjectively, non-embodied, disembodied, disbodied, discarnate, or bodiless, human consciousnesses of the dream-sleep-state.

Thus, both items, namely, the non-physical, non-material, unphysical, or immaterial, 3-D or three-dimensional, dreamal-cosmic-space of the dream-sleep-state on one hand, and the physical-looking, 3-D or three-dimensional, dreamal-cosmos or subjective-cosmos of the dream-sleep-state on the other, are situated or seated inside the subjectively, non-embodied, disembodied, disbodied, discarnate, or bodiless human consciousnesses of the dream-sleep-state.

Both the items mentioned above, namely, the non-physical, non-material, unphysical, or immaterial, **3-D** or **three-dimensional, dreamal-cosmic-space** of the dream-sleep-state on one hand, and the **physical-looking, 3-D** or **three-dimensional, dreamal-cosmos** or **subjective-cosmos** of the dream-sleep-state on the other, do not look, seem or appear at all or in the least different from the **3-D** or **three-dimensional cosmic space** as well as from the **3-D** or **three-dimensional physical cosmos** of the fully-awake or wide-awake-state.

To top it all, or the icing on the cake is the fact that the **3-D** or **three-dimensional, dreamal-cosmic-space** of the dream-sleep-state on one hand, and the **3-D** or **three-dimensional, physical-looking, dreamal-cosmos** or **subjective-cosmos** of the dream-sleep-state on the other, are both regarded to be absolutely real or one hundred percent genuine by all the subjectively, non-embodied, disembodied, disbodied, discarnate, or bodiless human consciousnesses of the dream-sleep-state, just as the **3-D** or **three-dimensional cosmic space** of the fully awake or wide awake state on one hand and the **3-D** or **three-dimensional physical-cosmos** or **objective-cosmos** of the fully awake or wide awake state on the other, are both regarded to be absolutely real or one hundred percent genuine by all the physically-embodied or objectively-embodied human consciousness of the fully awake or wide awake state.

But the best of all, is the fact that, every now and then, during the dream-sleep-state, the subjectively, non-embodied, disembodied, disbodied, discarnate, or bodiless, observer, seer, perceiver, beholder, or spectator of the **physical-looking** or **material-looking, dreamal-cosmos** or **subjective-cosmos** of the dream-sleep-state, observes, sees, perceives, or beholds, a twin, double, carbon copy, replica, or, lookalike of itself inside its **physical-looking** or **material-looking, dreamal-cosmos** or **subjective-cosmos** of the dream-sleep-state. This twin or double of the perceiver or the observer, looks or appears to the perceiver or the observer, physically or materially **embodied** in the manner, the physically or materially **embodied** human consciousnesses of the **physical cosmos** of the **fully-**

awake or **wide-awake-state** are. **This** twin or double of the perceiver or observer joins in or gets involved in as a participant or player in the affairs of the **physical-looking** or **material-looking, dreamal-cosmos** or **subjective-cosmos** of the dream-sleep-state.

In the above -mentioned scenarios, setting, or situation, a human consciousness of the dream sleep state becomes split into two conscious beings. One is the perceiving or the observing-aspect of the consciousness, and the other is the participating-aspect of the consciousness. The perceiving or the observing-aspect of the consciousness is **disembodied,** at least subjectively, whereas the participating-aspect, looks or appears to the perceiving or the observing-aspect, physically or materially embodied.

The above described scenarios, setting, or situation, wherein the splitting of one consciousness into two takes place, is akin to the situation which prevails in the current **fully-awake** or **wide-awake-state-cosmos** wherein the supra-transcendental consciousness i.e. god, or the creator of the current cosmos aka non-physical, non-material, unphysical, or immaterial, ubiquitous, and infinite, **3-D** or **three-dimensional, field of consciousness** or **ocean of consciousness,** known as **cosmic-space,** is the observer, seer, or, the perceiver of the current **physical cosmos** as well as the participator, partaker, sojourner, vacationer, traveller, voyager, wayfarer, sightseer, or globetrotter in the current **physical cosmos** in the form of all **physically** or **materially embodied conscious beings,** for example, all human beings.

The **physically-embodied** human consciousnesses of the **physical-cosmos** of mankind's **fully-awake** or **wide-awake-state,** must be reminded of the following fact in order that the above -mentioned truth with regards to the **physical-cosmos** of their **fully-awake** or **wide-awake-state** on one hand, and with regards to the **dreamal-cosmos** of their **dream-sleep-state** on the other, becomes permanently etched in their consciousnesses.

What has been said above can be put in another way.

The **objectively-embodied** human consciousnesses of the **objective-cosmos,** of mankind's **fully-awake** or **wide-awake-state,** must be reminded of the following fact, in order that the above-mentioned truth with regards to the **objective cosmos** of their **fully-awake** or **wide-awake-state** on one hand, and with regards to the **subjective-cosmos** of their **dream-sleep-state** on the other, becomes permanently etched in their consciousnesses.

The physically-embodied human consciousnesses of the physical-cosmos of mankind's fully-awake or wide-awake-state, must be reminded of the fact that, all the dreamal-events, dreamal phenomena, dreamal sights, or, dreamal spectacles, plus all the dreamal beings and things, which are seen, observed, perceived, and experienced by the subjectively-disembodied human consciousnesses of the dream-sleep-state with the aid of their dreamal senses, or, with the aid of their consciousnessbal-senses, appear to them as real, genuine, or authentic as all the physical-events, physical phenomena, physical sights, or, physical spectacles, plus all the physical beings and things, of their fully-awake or wide-awake-state-cosmos appear to their physical or material senses.

What has been said above can be put in another way.

The objectively embodied human consciousnesses of the objective cosmos of mankind's fully-awake or wide-awake-state must be reminded of the fact that all the subjective-events, subjective phenomena, subjective sights, or, subjective spectacles, plus all the subjective, beings and things, which are seen, observed, perceived, and experienced by the subjectively-disembodied human consciousnesses of the dream-sleep-state with the aid of their subjective-senses, or, with the aid of their consciousnessbal-senses, appear to them as real, genuine, or authentic as all the objective events, objective phenomena, objective sights, or, objective spectacles, plus all the objective beings and things of their fully-awake or wide-awake-state-cosmos appear to their objective senses.

The additional point for the human consciousnesses to take notice of, is that they see, observe, perceive and experience all the physical or, the objective events, phenomena, sights, or spectacles, plus beings and things of the physical or, the objective-cosmos in their fully-awake or wide-awake-state, while they are physically or, objectively embodied whereas they see, observe, perceive & experience all the dreamal or, the subjective events, phenomena, sights, or spectacles plus beings and things of their dreamal or, subjective cosmos of their dream-sleep-state, while they are dreamly or subjectively- disembodied.

The **physical cosmos** on one hand, and the non-physical, non-material, unphysical, or immaterial, ubiquitous, and infinite, **3-D** or **three-dimensional, field of consciousness** or **ocean of consciousness** aka **cosmic-space** aka supra-transcendental consciousness aka god, creator, or maker of the current cosmos on the other, which the **physically-embodied** or, better still, which the **objectively-embodied** human consciousnesses, observe, see, perceive, and experience in their **fully-awake** or **wide-awake-state** with the aid of their **physical** or **objective senses** and of which they are a **constituent-part,** plus in which they participate or engage throughout their lives, is a **daydream, reverie, imagery, dreamry,** or **fantasy** of the non-physical, non-material, unphysical, or immaterial, ubiquitous, and infinite, **3-D** or **three-dimensional, field of consciousness** or **ocean of consciousness** aka **cosmic-space** aka supra-transcendental consciousness aka god, creator, maker, or, progenitor of the current cosmos.

The **physically-embodied** or, **objectively-embodied,** human consciousnesses of the **fully-awake** or **wide-awake-state,** have been bequeathed the power to **daydream** by their **source** or **fountainhead** namely, the non-physical, non-material, unphysical, or immaterial, ubiquitous, and infinite, **3-D** or **three-dimensional, field of consciousness** or **ocean of consciousness** aka **cosmic-space** aka supra-transcendental consciousness aka god, creator, maker, or, progenitor of the current cosmos, so that the **physically-embodied** or, **objectively-embodied** human consciousnesses of the **fully-awake** or **wide-awake-state,** can also **daydream** as and when they want, desire, or feel like doing so and thus, create or give birth to their own, personal, or, private **subjective-cosmos, dreamal-cosmos** or **daydream-stuff-composed cosmos** of their fancy, liking, taste, or, preference, inside their own consciousnesses.

The **physically-embodied** or, better still, **objectively-embodied** human consciousnesses of the **fully-awake** or **wide-awake-state,** have been bequeathed the power to **daydream** by their creator, maker, or progenitor namely, the non-physical, non-material, unphysical, or immaterial, ubiquitous, and infinite, **3-D** or **three-dimensional, field of consciousness** or **ocean of consciousness** aka **cosmic-space** aka supra-transcendental consciousness aka god, creator, maker, or progenitor of the current cosmos, in order that someday they may realise that, the current **physical cosmos** or, the **objective-cosmos,** of which they are a **constituent-part,** has been made, created or formed by the very same method of **daydreaming,** by the non-physical, non-material, unphysical, or immaterial, ubiquitous, and infinite, **3-D** or **three-dimensional, field of consciousness** or **ocean of consciousness** aka **cosmic-space** aka supra-transcendental consciousness aka god, creator, maker, or progenitor of the current cosmos.

To sum up.

The current 3-D or three-dimensional, physical or, objective cosmos of mankind's fully-awake or, wide-awake-state, which consists of countless, physical or, objective, events, phenomena, sights, or, spectacles, plus, beings and things, as well as, moons, planets, stars, galaxies, and the like, is floating, wafting, or levitating, plus, whirling, twirling or spiralling non-stop from the beginning of the current time as well as from the beginning of the current physical or objective cosmos, inside the non-physical, non-material, unphysical, or immaterial, ubiquitous, and infinite, 3-D or three-dimensional, field of consciousness or ocean of consciousness aka cosmic-space aka supra-transcendental consciousness aka god, creator, maker, or progenitor of the current cosmos, as a mere daydream, reverie, imagery, dreamry, or, fantasy of this non-physical, non-material, unphysical, or immaterial, ubiquitous, and infinite, 3-D or three-dimensional, field of consciousness or ocean of consciousness aka cosmic-space aka supra-transcendental consciousness aka god, creator, maker, or progenitor of the current cosmos, nothing more nothing less, and the physical or objective bodies of the human consciousnesses of the current cosmos are merely a constituent-part of this daydream, reverie, imagery, dreamry, or, fantasy of the non-physical, non-material, unphysical, or immaterial, ubiquitous, and infinite, 3-D or three-dimensional, field of consciousness or ocean of consciousness aka cosmic-space aka supra-transcendental consciousness aka god, creator, maker or progenitor of the current cosmos, nothing more nothing less.

Thus, the physical or objective cosmos is also a dreamal or subjective cosmos but dreamal or subjective cosmos in relation to or vis-a-vis its god, creator, maker, or progenitor who is none other than the non-physical, non-material, unphysical, or immaterial, ubiquitous and infinite, 3-D or three-dimensional field of consciousness or ocean of consciousness aka cosmic space which, in turn, is the expanded, distended, dilated, or inflated form or version of the dimensionless form or version, or, the original form or version of the supra-transcendental consciousness aka god, creator, maker, or progenitor of the current physical or objective cosmos of mankind's fully-awake or wide-awake-state.

THE PHYSICAL COSMOS IS A QUALIFIED REALITY AND NOT A DEFINITIVE REALITY

The view point which most human consciousnesses hold in their fully awake or wide awake state is that their **'dream-sleep-state-cosmos'** is **'unreal'** (because it is only a **dream**) and their **'fully-awake'** or **'wide-awake-state-cosmos'** is **'real'** (because it can not only be seen by one's physical eyes but can also be stamped by one's physical feet, touched by one physical fingers, smelt through one's physical nose, tasted by one's physical tongue and heard through one's physical ears, a viewpoint which all of them defend very forcefully in their fully awake or wide awake state on every possible occasion), is an utterly fallacious view point because it is based on an extreme degree of intellectual short-sightedness on their part.

The opinion regarding the 'reality' or 'unreality' of an observed or perceived phenomenon, held by an observing or perceiving human consciousness, is entirely dependent upon the context or the surrounding within which the observed or the perceived phenomenon and the observing or perceiving human consciousness find themselves together.

That is to say, the idea of 'reality' or 'unreality' which an observing or perceiving human consciousness holds with regards to an observed or perceived phenomenon is always merely a relative, comparative, or qualified idea and never an absolute, conclusive, or definitive idea because it is always totally dependent upon the circumstance or the environment in which the phenomenon which is being observed or perceived and the consciousness who observes or perceives that phenomenon find themselves together.

The definition of 'what constitutes reality' changes from environment to environment, from backdrop to backdrop, or from setting to setting in which a phenomenon which is being observed or perceived and a human consciousness who observes or perceives that phenomenon find themselves together or in which an 'object' which is being observed or perceived and a 'subject' who observes or perceives that 'object' find themselves together.

God, or the creator of the current physical cosmos aka **supra-transcendental consciousness** is the one and only absolute, final, or conclusive truth there is at any instant, anywhere i.e. both during the existence of the current time and current physical cosmos as well as after the ending of the current time and current physical cosmos.

The entire god-created or god-manufactured physical cosmos including all its 'objects', events, phenomena, sights, or spectacles are relative, comparative, or qualified truths only and not the absolute, ultimate, definitive, or final truth in the manner of god, creator, or maker of the current cosmos.

Now, let's change the theme a little bit.

All the **'subjects'** or all the **'individual-consciousnesses'** of the current physical cosmos (for example, all the human consciousnesses of the current physical cosmos) are in the **physically embodied form.** It is they who actually reside within this god-created or god-manufactured current physical cosmos in a **physically embodied form.** Therefore, it is they and only they who observe, perceive, and experience all the physical 'objects', events, phenomena, sights, or spectacles of this physical cosmos from within, both as an active participant or player and as a passive observer or spectator.

Even though the current **physical cosmos** i.e. the **physical matter** of the current cosmos – (which includes the **physical bodies** of all the **'subjects'** or all the **'individual-consciousnesses'** of the current **physical cosmos**) - itself is **transient, temporary,** or **time-bound,** these **'subjects'** or these **'individual-consciousnesses'** of the current **physical cosmos** themselves (for example, the **human consciousnesses** of the current **physical**

cosmos themselves) are **timeless, immortal,** or **eternal** in the manner of god, creator, or maker of the current **physical cosmos,** because they all, without exception, are absolutely **unmodified, unaltered, unvitiated** or, **pristine** or **pure** and **uncorrupted** form or version of a section, segment, part, or portion of the ubiquitous & infinite **field of consciousness** or **ocean of consciousness** aka **cosmic space** which, in turn, is nothing but the **expanded, distended, dilated,** or **inflated** form or version of the **dimensionless** form or version, that is to say, of the **original** form or version of the **timeless, immortal,** or **eternal** god, creator or maker of the current cosmos. Inside this **expanded, distended, dilated,** or **inflated** form or version of the **dimensionless** form or version, or, the **original** form or version of the **timeless, immortal,** or **eternal** god, creator or maker of the current cosmos aka **cosmic space,** the current **physical cosmos,** that is to say, the current **physical-matter-composed,** all objects, things, items, or entities of the current cosmos, for example, the current **physical-matter-composed,** all moons, planets, stars, galaxies and the like of the current cosmos are **floating, wafting,** or **levitating** plus **whirling, twirling,** or **spiralling** non-stop from the very **beginning** of the **current time** and the **current physical cosmos** and will keep on doing so till the very **end** of the **current time** and the **current physical cosmos,** as per the wish, will, whim, or fancy of god, creator, or maker of the current **physical cosmos** who is none other than the **non-physical,** ubiquitous and infinite **field of consciousness** or **ocean of consciousness** called **cosmic space.**

Incidentally, the **human idea** called **'time'** always was, always is, and always will be a **'relative'** or **'qualified'** truth. This **human idea** called **'time'** never was, never is, and never will be an **'absolute'** or **'eternal'** truth. This is so because this **human idea** of **'time'** is applicable and therefore, relevant only in respect of the **time-bound, transient,** or **temporary, physical cosmos** i.e. in respect of the **time-bound, transient,** or **temporary, physical matter** of the current cosmos. That is to say, it is applicable and therefore, relevant only in respect of the **physical-matter-composed** objects, things, items, or entities of the current cosmos, for example, the **physical-matter-composed** bodies of all the **physically embodied** consciousnesses of the current cosmos and all the **physical-matter-composed** insentient or incapable of feeling or understanding things, objects, or entities such as moons, planets, stars, galaxies and the like of the current **physical cosmos.** It is not applicable and therefore, irrelevant in respect of all the **consciousnesses** of the current cosmos, irrespective of whether the **consciousness** in question is the **timeless, immortal,** or **eternal consciousness** of god, creator, or maker of the current **physical cosmos** i.e. the current **cosmic space,** or they are the **timeless, immortal,** or **eternal consciousnesses** who inhabit, abide, or exist inside the current cosmos in **physically embodied** form, for example, all the **physically embodied,** human consciousnesses of the current cosmos.

The label or the epithet 'relative reality' is correct only with regards to all the 'objects' or, more to the point, with regards to all the 'physical objects' of the current physical cosmos and not with regards to all the 'subjects' or, more to the point, and not with regards to all the 'individual consciousnesses' of the current physical cosmos, for example, all the 'individual human consciousnesses' of the current physical cosmos.

The label or the epithet **'subject'** is pinned to all the individual **physically-embodied** consciousnesses of the current physical cosmos, for example, all the individual, **physically-embodied** human consciousnesses of the current cosmos because it is they who observe, perceive, and experience all the physical **'objects'** of the current physical cosmos.

Thus, a **'subject'** is a **physically-embodied** consciousness who personally observes, perceives, and experiences another entity that exists outside it (called an **'object'**), or who has a relationship with another entity that exists outside it (called an **'object'**).

In other words, a **'subject'** is an observer and an **'object'** is a thing observed.

In the realm or domain of Adwait-Vedanta, the label or the epithet 'prapancha', 'phenomenon', 'relative-reality', 'comparative-reality', or 'qualified-reality' is pinned or fastened both to the fully-awake-state or wide-awake-state-physical-cosmos as well as to all the dream-sleep-state-dreamal-cosmoses. All the dream-sleep-state-dreamal-cosmoses seem, look, or appear to all the human consciousnesses of the dream-sleep-state, exactly, absolutely, or one hundred percent like the fully-awake-state or wide-awake-state-physical-cosmos.

The human consciousnesses who observe, perceive, and experience the above mentioned two cosmoses (or, better still, the above mentioned two **'prapanchas', 'phenomena', 'relative-realities', 'comparative-realities',** or **'qualified-realities'**) i.e. the **physical cosmos** of their fully-awake or wide-awake state on one hand and the **physical-looking, dreamal-cosmoses** of their dream-sleep-state on the other, are called **'subjects'** as a counterpoint to the word **'object'** which is applied to all the **'prapanchas', 'phenomena', 'relative-realities', 'comparative-realities',** or **'qualified-realities'** such as the **fully-awake-state** or **wide-awake-state-physical-cosmos** on one hand and all the **dream-sleep-state-dreamal-cosmoses** on the other.

What has been said above can be put in another way.

Both kinds of cosmoses i.e. the physical cosmos of the fully-awake or wide-awake-state of human consciousnesses on one hand and the physical-looking, **dreamal-cosmoses** of the dream-sleep-state of human consciousnesses on the other, are 'objects' , 'prapanchas', 'phenomena' , 'relative-realities , 'comparative-realities', or 'qualified-realities' only, nothing more nothing less, which are observed or perceived by the human senses, and then presented to human consciousnesses for interpretation and deriving experiences therefrom.

That is to say, they both are 'objects', 'prapanchas', 'phenomena', 'relative-realities, 'comparative-realities', or 'qualified-realities' only in relation to their observing, perceiving, and experiencing human consciousnesses namely the 'subjects', nothing more nothing less.

To reiterate.

Both kinds of cosmoses aka 'objects', 'prapanchas', 'phenomena', relative-realities, 'comparative-realities', or 'qualified-realities' have their own human 'subjects' i.e. human observers, perceivers, and experiencers, or, if it is preferred, human consciousnesses who observe, perceive and experience their respective cosmoses aka their respective 'objects', 'prapanchas', 'phenomena', 'relative-realities, 'comparative-realities', or 'qualified-realities', and try to make sense of them plus try to react with them or deal with them in the manner they deem fit at the time. They finally declare them to be real or unreal as per their understanding, from the perspective of the surroundings, circumstances, or conditions they find themselves.

Thus, the observing, perceiving, and experiencing human consciousnesses i.e. the 'subjects', who observe, perceive, and experience the 'object', 'prapancha', 'phenomenon' , 'relative-reality' , 'comparative-reality', or 'qualified-reality' called the **fully-awake** or **wide-awake-state-physical-cosmos,** do not have any right to say that it is only that 'object' , 'prapancha', 'phenomenon' , 'relative-reality', comparative-reality', or 'qualified-reality' , which they observe, perceive, and experience during their **fully-awake** or **wide-awake-state (namely fully-awake** or **wide-awake-state-physical-cosmos)** which is real whereas the other 'object', 'prapancha', 'phenomenon','relative-reality', 'comparative-reality', or 'qualified-reality' called the **dream-**

sleep-state-dreamal-cosmos is unreal.

The absolute truth is that both these cosmoses, 'objects', 'prapanchas', 'phenomena', 'relative-realities, 'comparative-realities', or 'qualified-realities' namely, the **fully-awake** or **wide-awake-state-physical-cosmos** on one hand and the **dream-sleep-state-dreamal-cosmoses** on the other, plus all their 'objects', 'prapanchas', or 'phenomena' are relative, comparative, or qualified truths only, nothing more nothing less. None of them is an 'absolute' or 'final' truth. The label or the epithet of 'absolute' or 'final' truth belongs to god only, or belongs to the creator of these two cosmoses, 'objects', 'prapanchas', 'phenomena', 'relative-realities', 'comparative-realities', or 'qualified-realities' only, and to no one else.

Hence, the human consciousnesses of the **fully-awake** or **wide-awake-state-physical-cosmos** must not harbour a hang up, neurosis, obsession, or mental block, or, better still, an absolutist viewpoint with regards to the 'reality' of the **fully-awake** or **wide-awake-state-physical-cosmos** on one hand and with regards to the 'unreality' of their **dream-sleep-state-dreamal-cosmoses** on the other. It must be acknowledged by them that the **fully-awake** or **wide-awake-state-physical-cosmos** is also a mere 'object', 'prapancha', 'phenomenon', 'relative-reality', 'comparative-reality', or 'qualified-reality' in the manner their **dream-sleep-state-dreamal-cosmoses** are, no more no less.

It is not surprising that the human consciousnesses when they find themselves in their **dream-sleep-state,** they regard their **dream-sleep-state-dreamal-cosmoses** to be 'real' and when they find themselves in their **fully-awake** or **wide-awake-state,** they regard the **fully-awake** or **wide-awake-state-physical-cosmos** to be 'real'. This is due to the fact that both these **cosmoses** seem, look, or appear so 'real' to human consciousnesses in their respective milieu, habitat, domain, or turf.

In the realm of Adwaita-Vedanta, the Sanskrit word 'maya' is used to designate the delusion, misapprehension, or misconception, inserted or installed by god into the human consciousnesses so that they may take part in the god-created or god-manufactured 'prapancha', 'phenomenon' , 'relative-reality' , 'comparative-reality', or 'qualified-reality' called the **fully-awake** or **wide-awake-state-physical-cosmos** with full effort, energy, and enthusiasm, by thinking, believing, or imagining it to be one hundred percent real, which it is nothing of the sort for it is merely a 'prapancha', 'phenomenon', 'relative-reality' , 'comparative-reality', or 'qualified-reality' on par with, neck and neck with, or matching with the **dream-sleep-state-dreamal-cosmos,** nothing more nothing less.

The Sanskrit word 'maya' is defined as a peculiar, off centre, or quirky belief or impression, maintained despite being contradicted by rational argument.

The Sanskrit term 'prapancha' is used in the realm of 'Adwait-Vedanta' to designate the god-created or god-manufactured 'phenomenon', 'relative-reality', 'comparative-reality', or 'qualified-reality' called the **fully-awake** or **wide-awake-state-physical-cosmos.**

The Sanskrit word 'prapancha' comes from the Sanskrit root 'pach', which means 'to spread out', 'proliferate', or 'elaborate'. Thus, this Sanskrit word 'prapancha' depicts, delineates, or describes the tendency of the consciousness to spread out, proliferate, or elaborate on the bare perception of the senses.

THE PHYSICAL COSMOS IS A DREAMAL COSMOS IN THE MANNER THE DREAM-SLEEP-STATE COSMOS IS

On account of their god-bequeathed, intellectual myopia, the great majority of the **fully-awake** or **wide-awake-state** human consciousnesses hold on steadfastly to their delusion or, wrong notion that it is only their **dream-sleep-state-dreamal-cosmoses** which are 'dreamal' in nature and therefore, 'unreal' and not their **fully-awake** or **wide-awake-state-physical-cosmos,** which they say, is one hundred percent 'real' because, it is not 'dreamal' in nature and instead is 'physical' in nature. The reason they put forward, for their **fully-awake** or **wide-awake-state-physical-cosmos** to be not 'dreamal' in nature and therefore, absolutely 'real', is that, they say, they can, not only see their **fully-awake** or **wide-awake-state-physical-cosmos** by their physical eyes, but they can also stamp on it with their physical feet, touch it by their physical fingers, smell it through their physical nose, taste it by their physical tongue, and hear it through their physical ears. This kind of powerful justification by the human consciousnesses of the **fully-awake** or **wide-awake-state** about the absolute 'reality' of their **fully awake** or **wide-awake-state-physical cosmos** is on account of the strong hold, which the dominance or supremacy of god-bequeathed **maya** or **misapprehension** has on the human consciousnesses. This is the god-bequeathed **magic of maya** or, **magic of misapprehension** at its best or, at its highest standard of quality possible. This god-bequeathed **magic of maya** or, **magic of misapprehension** has bewitched the **fully-awake** or **wide-awake-state** human consciousnesses from the time of advent of man in the current cosmos.

The above mentioned, god-bequeathed, intellectual-tunnel-vision, afforded to the **fully-awake** or **wide-awake-state** human consciousnesses, is the cause or the reason due to which the **fully-awake** or **wide-awake-state** human consciousnesses are not able to work out the absolute truth regarding the **true source** or **fountainhead** of their **fully-awake** or **wide-awake-state-physical-cosmos.**

Moreover, the above mentioned, god-bequeathed, intellectual-short-sightedness, forces, coerces, or compels the **fully-awake** or **wide-awake-state** human consciousnesses to believe, think, or imagine that the god-manufactured or god-created 'prapancha', 'phenomenon' , 'relative-reality' , 'comparative-reality', 'qualified-reality', or, 'dreamal' reality called the **physical-matter/physical energy duo** (which composes the current **fully-awake** or **wide-awake-state-physical-cosmos)** has come out of, or, if it is preferred, has originated, emerged, or emanated from an absolutely insentient or incapable of feeling or understanding, **infinitesimally small, infinitely hot,** and **infinitely dense** lump of **3-D** or **three-dimensional, physical matter** called **singularity** or **cosmic egg.**

Additionally, the above mentioned god-bequeathed, intellectual-tunnel-vision, forces, coerces, or compels the **fully-awake** or **wide-awake-state** human consciousnesses to believe, think, or imagine that the current **nonphysical, nonmaterial, unphysical,** or **immaterial,** ubiquitous and infinite **cosmic space,** and all the current **nonphysical, nonmaterial, unphysical,** or **immaterial consciousnesses** which inhabit, abide, or exist in the **physically embodied** or **materially embodied** form - (for example, all the **physically** or **materially embodied,** human consciousnesses of the current cosmos) - inside the current **nonphysical, nonmaterial, unphysical,** or **immaterial,** ubiquitous and infinite, **cosmic space**, have come out of, or, if it is preferred, have originated, emerged, or emanated from the above mentioned, absolutely insentient or incapable of feeling or understanding, **infinitesimally small, infinitely hot,** and **infinitely dense** lump of **3-D** or **three-dimensional, physical matter** called **singularity** or **cosmic egg.**

But the truth is quite the opposite. And this truth is discussed below.

God, creator, or maker of the current physical cosmos is the current nonphysical, nonmaterial, unphysical, or immaterial, ubiquitous and infinite, 3-D or three-dimensional, cosmic space which is nothing but an awe-inspiring field of consciousness or ocean of consciousness because it is the expanded, distended, dilated, or inflated form or version of the dimensionless form or version, or, the original form or version of god, creator, or maker of the current physical cosmos.

All the current **nonphysical, nonmaterial, unphysical,** or **immaterial, consciousnesses** of the current cosmos, for example, all the human consciousnesses of the current cosmos, which inhabit, abide, or exist in **physically** or **materially embodied** form inside the current **nonphysical, nonmaterial, unphysical,** or **immaterial,** ubiquitous and infinite, **field of consciousness** or **ocean of consciousness** aka **cosmic space,** are absolutely **unmodified, unaltered, unvitiated** or, **pristine** or **pure** and **uncorrupted** form or version of a section, segment, part, or portion of this **nonphysical, nonmaterial, unphysical,** or **immaterial,** ubiquitous and infinite **field of consciousness** or **ocean of consciousness** aka **cosmic space** which, as said before, is nothing but the **expanded, distended, dilated,** or **inflated** form or version of the **dimensionless** form or version, or, the **original** form or version of god, creator, or maker of the current **physical cosmos.**

God, creator, or maker of the current cosmos existed in its **original** form or version or, **dimensionless** form or version some 13.7 billion light years ago, before the beginning of the **current time,** and beginning of the current **cosmic space** plus the beginning of the current **physical cosmos.** This god, creator, or maker of the current cosmos now exists as **nonphysical, nonmaterial, unphysical,** or **immaterial,** ubiquitous and infinite, **3-D** or **three-dimensional, field of consciousness** or **ocean of consciousness** aka **cosmic space** which, as said before, is nothing but the **expanded, distended, dilated,** or **inflated** form or version of the **dimensionless** form or version, or, the **original** form or version of god, creator, or maker of the current **physical cosmos.**

The **original** form or version of god, creator, or maker of the current cosmos, which existed some 13.7 billion light years ago, before the beginning of the **current time** and the beginning of the current **cosmic space** plus the beginning of the current **physical cosmos,** is described as the incredible or awe-inspiring and **the only one of its kind,** or unique, **nonphysical, nonmaterial, unphysical,** or **immaterial, timeless, bodiless,** and **dimensionless consciousness** or **awareness** of infinite intelligence, imagination, and emotion.

While on the subject, let one inform the human consciousnesses of the **fully-awake** or **wide-awake-state** that the **3-D** or **three-dimensional, physical matter/physical energy duo** of the current cosmos, has been has been put together or produced, or, if it is preferred, has been constructed or created, out of a condensed, compressed or compacted or, if it is preferred, out of a consolidated or congealed or, better still, out of a concretized or solidified section, segment, part or portion of the **nonphysical, nonmaterial, unphysical,** or **immaterial,** ubiquitous and infinite, **3-D** or **three-dimensional, field of consciousness** or **ocean of consciousness** aka **cosmic space** and not out of an absolutely unmodified, unaltered, unvitiated or, absolutely pristine, or pure and uncorrupted form or version of a section, segment, part, or portion of the **nonphysical, nonmaterial, unphysical,** or **immaterial,** ubiquitous and infinite, **3-D** or **three-dimensional, field of consciousness** or **ocean of consciousness** aka **cosmic space** as is the case with all the current **nonphysical, nonmaterial, unphysical,** or **immaterial, consciousnesses** of the current cosmos, for example, all the human consciousnesses of the current cosmos, which inhabit, abide, or exist in the **physically** or **materially embodied** form inside the current **nonphysical, nonmaterial, unphysical,** or **immaterial,** ubiquitous and infinite, **3-D** or **three-**

dimensional, field of consciousness or **ocean of consciousness** aka **cosmic space** which is nothing but the expanded, distended, dilated, or inflated form or version of the **dimensionless** form or version, or, the **original** form or version of god, creator, or maker of the current cosmos.

Just to remind one that the **3-D** or **three-dimensional, physical matter/physical energy duo** of the current cosmos is the thing or substance which has composed, constituted, or, formed the **physical** or the **material bodies** of all the **embodied** consciousnesses of the current cosmos, for example, the **physical** or the **material bodies** of all the **embodied** human consciousnesses of the current cosmos and which also has composed, comprised, constituted, or, formed all the absolutely insentient or incapable of feeling or understanding things, items, or entities of the current cosmos such as all the moons, planets, stars, galaxies and the like of the current cosmos. All the moons, planets, stars, galaxies and the like of the current cosmos are the things, items, or entities, which are **floating, wafting,** or **levitating** plus **whirling, twirling,** or **spiralling** non-stop inside the **nonphysical, nonmaterial, unphysical,** or **immaterial,** ubiquitous and infinite, **3-D** or **three-dimensional, field of consciousness** or, **ocean of consciousness** called **cosmic space** which is nothing but the expanded, distended**, dilated,** or **inflated** form or version of the **dimensionless** form or version, or, the **original** form or version of god, creator, or maker of the current cosmos.

Now, let's change the theme of discussion a little bit so that, the point at issue, or, better still, so that, the bone of contention at hand, namely that science's, absolutely insentient or incapable of feeling or understanding, **infinitesimally small, infinitely hot,** and **infinitely dense,** lump of **3-D** or **three-dimensional,** physical matter, called **singularity** or **cosmic egg,** of Big Bang theory fame, cannot be the source, or, fountainhead of the current sentience-filled, enlightenment-filled, consciousnesses-filled, awarenesses-filled, or, mind-filled cosmos, can be discussed, debated, examined, explored, analysed, or, scrutinised, in great depth or detail.

That is to say, so that, the point at issue, or, the bone of contention at hand, namely that, the science's, absolutely insentient or incapable of feeling or understanding, **infinitesimally** tiny bit of **physical matter,** which is **picturesquely** described, by the Big Bang theory of science, as the **infinitesimally small, infinitely hot,** & **infinitely dense, 3-D** or **three-dimensional,** physical or material thing, or, entity, called **singularity** or **cosmic egg,** cannot be the source, or, fountainhead of the **nonphysical, nonmaterial, unphysical,** or **immaterial,** ubiquitous and infinite **space,** or, more to the point, **cosmic space** of the current cosmos on one hand, and all the **nonphysical, nonmaterial, unphysical,** or **immaterial,** sentience, enlightenment, consciousnesses, awarenesses, or minds of the current cosmos, for example, all the human sentience, enlightenment, consciousnesses, awarenesses, or minds of the current cosmos, on the other, can be discussed debated, examined, explored, analysed, or, scrutinised, in great depth or detail.

What has been said above can be put in another way.

The point at issue, or, better still, the bone of contention, at hand is**: - "Can sentience be borne out of insentience** i.e. can enlightenment, consciousnesses, awarenesses, or, minds be born out of absolutely insentient or incapable of feeling or understanding, physical matter or, is the **supreme**

truth, the other way around, that, the **sentence is the mother of both sentience and insentience** i.e. that the enlightenment, consciousness, awarenesses, or, mind aka god aka cosmic space is the mother of both enlightenment, consciousnesses, awarenesses, or, minds as well as physical matter?

In order to enlarge further on the theme, dealt with, immediately above, one will now like to continue, in the following manner.

The **current cosmos** consists of three ingredients:-

Firstly, the **nonphysical, nonmaterial, unphysical,** or **immaterial,** ubiquitous and infinite **space** or, more to the point, **cosmic space** of the current cosmos.

Secondly, all the **nonphysical, nonmaterial, unphysical,** or **immaterial,** physically or materially, **embodied** consciousnesses of the current cosmos, for example, all the human consciousnesses of the current cosmos.

Thirdly, the **3-D** or **three-dimensional, physical matter/physical energy duo** of the current cosmos which makes or composes all the insentient, or, all the incapable of feeling or understanding things, objects, or, entities of the current cosmos, for example, all the moons, planets, stars, galaxies and the like of the current cosmos on one hand and the physical bodies of all the physically **embodied** consciousnesses of the current cosmos on the other, for example, the physical bodies of all the human consciousnesses of the current cosmos.

As said before, the **3-D** or **three-dimensional, physical**

matter/physical energy duo of the current cosmos is the thing which has composed the **physical** or **material bodies** of all the **embodied consciousnesses** of the current cosmos, for example, the **physical** or **material bodies** of all the human consciousnesses of the current cosmos. This **3-D** or **three-dimensional, physical matter/physical energy duo** of the current cosmos also has composed all the totally **insentient** or totally incapable of feeling or understanding things, items, or entities of the current cosmos such as all the moons, planets, stars, galaxies and the like of the current cosmos. All the moons, planets, stars, galaxies and the like of the current cosmos are the items of the current cosmos which are **floating, wafting,** or **levitating** plus **whirling, twirling,** or **spiralling** non-stop inside the **nonphysical, nonmaterial, unphysical,** or **immaterial,** ubiquitous and infinite, **3-D** or **three-dimensional space** or, more to the point, **3-D** or **three-dimensional cosmic space** of the current cosmos.

The scientific theory of Big Bang states that the **current cosmos** consisting of the above three ingredients (namely, the current **nonphysical, nonmaterial, unphysical,** or **immaterial,** ubiquitous and infinite **cosmic space** , all the current **nonphysical, nonmaterial, unphysical,** or **immaterial,** physically or materially **embodied** consciousnesses of the current cosmos, for example, all the human consciousnesses of the current cosmos and the current **physical matter/physical energy duo)** has emerged or, emanated from or, better still, has issued forth, or, has come out of an infinitesimally tiny bit of **insentient** or incapable of feeling or understanding, **physical matter** called **singularity** or **cosmic egg** of Big Bang theory fame. The full description of this infinitesimally tiny bit of **insentient** or incapable of feeling or understanding, **physical matter** called

singularity or **cosmic egg,** as given by the scientific theory of Big Bang, is that it is an **infinitesimally small, infinitely hot,** and **infinitely dense,** lump of **insentient** or incapable of feeling or understanding **physical matter.**

The premise, logic, or, proposition of the scientific theory of Big Bang, that the physical matter, albeit an infinitesimally small, infinitely hot and infinitely dense lump of insentient physical matter, is the source or the fountainhead of the current cosmos, seems quite off the mark. Why this is so is described below in detail but before that, let one describe, once again, what the physical matter of the current cosmos is and what its role in the current cosmos is.

The current **physical matter/physical energy duo** composes the **physical** or the **material bodies** of all the **embodied consciousnesses** of the current cosmos, for example, the **physical** or the **material bodies** of all the human consciousnesses of the current cosmos. It also composes all the insentient or, incapable of feeling or understanding, things, items, or entities of the current cosmos, for example, all the moons, planets, stars, galaxies and the like of the current cosmos, all of which are **floating, wafting,** or **levitating** plus **whirling, twirling,** or **spiralling** non-stop, inside the **nonphysical, nonmaterial, unphysical,** or **immaterial,** ubiquitous and infinite **space** or, more to the point **cosmic space.**

The reason why the above-mentioned premise, logic, or, proposition of the scientific theory of Big Bang, that the **physical matter,** albeit an **infinitesimally small, infinitely hot** and **infinitely dense** bit of **physical matter** is the **source** of the **current cosmos,** seems quite off the mark is explained in detail below.

Firstly, there is no evidence in the current cosmos that the **physical matter,** no matter how small in dimension or size, (i.e. even if it is of an infinitesimally small dimension or size, as the Big Bang theory states that its physical **singularity** or **cosmic egg** is), can ever be **dimensionless** so that it can be imagined, visualized, pictured, or conceptualized to exist without the need of any kind of pre-existent space or, more to the point, any kind of pre-existent **cosmic space,** that is to say, so that it can be imagined, visualized, pictured, or conceptualized to exist in the absence of any kind of pre-existent space or, more to the point, any kind of pre-existent **cosmic space,** in order that it may become reasonable, acceptable, believable, or, imaginable, or, better still, so that, it may become conceivable, thinkable, feasible, or, credible, for one to accept that the **source,** or the **fountainhead** of the current **nonphysical, nonmaterial, unphysical,** or **immaterial,** ubiquitous and infinite **cosmic space** is **physical matter,** or, more to the point, is the **infinitesimally small, infinitely hot** and **infinitely dense** bit of **physical matter** called **singularity** or **cosmic egg** of Big Bang theory fame.

Therefore, one feels that the premise, logic, or, proposition of the scientific theory of Big Bang, that the insentient or incapable of feeling or understanding, one hundred percent, **physical matter,** or, more to the point, the insentient or incapable of feeling and understanding, one hundred percent, **infinitesimally small, infinitely hot** and **infinitely dense** bit of **physical matter** called **singularity** or **cosmic egg** of Big Bang theory fame is the **source** or **fountainhead** of the current **nonphysical, nonmaterial, unphysical,** or **immaterial,** ubiquitous and infinite **cosmic space,** is quite off the mark.

Secondly, there is no evidence in the current cosmos of totally insentient, or, totally incapable of feeling or understanding, one hundred percent, **physical matter** giving birth to any kind or size of **nonphysical, nonmaterial, unphysical,** or **immaterial** space, let alone the current **nonphysical, nonmaterial, unphysical,** or **immaterial,** ubiquitous and infinite **cosmic space.**

Thirdly, there is no evidence in the current cosmos of totally insentient, or, totally incapable of feeling or understanding, one hundred percent, **physical matter** giving birth to any kind of **nonphysical, nonmaterial, unphysical,** or **immaterial** consciousness, for example, animal consciousness, let alone human consciousness.

Therefore, the premise, logic, or, proposition of the scientific theory of Big Bang that the insentient or incapable of feeling and understanding, one hundred percent, **physical matter,** or, more to the point, the insentient or incapable of feeling and understanding, one hundred percent, **infinitesimally small, infinitely hot** and **infinitely dense** bit of **physical matter** called **singularity** or **cosmic egg** of Big Bang theory fame is the **source** or **fountainhead** of the current **nonphysical, nonmaterial, unphysical,** or **immaterial,** ubiquitous and infinite **cosmic space** on one hand and of the current **nonphysical, nonmaterial, unphysical,** or **immaterial,** human and animal consciousnesses on the other, is quite off the mark. On the other hand, there is ample evidence in the current cosmos of **nonphysical, nonmaterial, unphysical,** or **immaterial,** consciousnesses giving birth, not only to other **nonphysical, nonmaterial, unphysical,** or **immaterial** consciousnesses but also to **nonphysical, nonmaterial, unphysical,** or **immaterial,** space, or, if it is preferred, cosmic space.

Furthermore, there is also ample evidence in the current cosmos of **nonphysical, nonmaterial, unphysical,** or **immaterial,** consciousnesses giving birth to insentient or incapable of feeling or understanding physical matter.

For example, each night, human consciousnesses in their dream-sleep state, create not only the **nonphysical, nonmaterial, unphysical,** or **immaterial** space i.e, **consciousnessbal space** inside their consciousnesses but also create totally insentient, or, totally incapable of feeling and understanding **physical matter** during their dream-sleep state.

For instance, each night, human consciousnesses in their dream-sleep state, create inside themselves, totally insentient, or, totally incapable of feeling or understanding, mountains, rivers, planets, stars, galaxies and the like. They also create inside themselves each night during their **dream-sleep state,** physically **embodied** consciousnesses, for example, physically **embodied** human consciousnesses and animal consciousnesses.

Therefore, the **theory of god** i.e. the **theory** that an incredible or awe-inspiring, and **the only one of its kind** or unique, **timeless, bodiless,** and **dimensionless consciousness** of infinite intelligence, imagination, and emotion called god, creator, maker, or, whatever is more likely to be true **source** or **fountainhead** of the current cosmos than the **singularity** or **cosmic egg** of Big Bang theory fame, that is to say, than an absolutely insentient or incapable of feeling or understanding, **infinitesimally tiny lump of** physical matter called **singularity** or, **cosmic egg,** or, if it is preferred, than an absolutely insentient or incapable of feeling or understanding, **infinitesimally small, infinitely hot** and **infinitely dense** bit of **physical matter** called **singularity** or **cosmic egg** of Big Bang theory fame.

DREAM-DREAMER AND DREAMER-DREAM RELATIONSHIP OR OBJECT-SUBJECT AND SUBJECT-OBJECT RELATIONSHIP IN THE CURRENT COSMOS

Human consciousnesses in their **fully-awake** or **wide-awake-state** must take into consideration the following fact about the **dreamal-cosmoses** or the **subjective-cosmoses** of their **dream-sleep-state** if they really wish to know what is the **true nature** of the **physical-cosmos** or the **objective-cosmos** which they see, perceive and experience plus in which they take part in their **fully-awake** or **wide-awake-state?**

Human consciousnesses, when they are in their **dream-sleep-state,** do not remember at all about the existence of the **physical-cosmos** or the **objective-cosmos** of their **fully-awake** or **wide-awake-state.** As a result, they are not in the least aware or conscious of the existence of this cosmos when they are in their **dream-sleep-state.**

In other words, from the perspective of the human consciousnesses of the **dream-sleep-state**, the **physical-cosmos** or the **objective-cosmos** of mankind's **fully-awake** or **wide-awake-state** is non-existent, not present, absent, without foundation, or null and void. As far as they are concerned, apart from the reality of their own existence as an **aware** or **conscious** and **physically** or **objectively, non-embodied, disembodied, disbodied, discarnate** or **bodiless-being,** the only other reality whose existence they are aware or conscious of in their **dream-sleep-state,** is that of the **physical-looking** or **objective-looking, dreamal-cosmos** or the **subjective cosmos** of their **dream-sleep-state** whom they regard at the time, as one hundred percent **physical** or **objective** and one hundred percent **real** or **authentic** and not at all **dreamal** or **subjective** and **unreal** or **unauthentic** as they do when they are in their **fully-awake** or **wide-awake-state** or, better still, as they do from the perspective of their **fully-awake** or **wide-awake-state.**

From the perspective of the human consciousnesses of the **dream-sleep-state,** the only **conscious beings** in existence for the purpose of interaction during their **dream-sleep-state** are the **dreamly-embodied** or **subjectively-embodied consciousnesses** of the **physical-looking** or **objective-looking, dreamal-cosmos** or the **subjective-cosmos** of their **dream-sleep-state,** for example, **dreamly-embodied** or **subjectively-embodied** human consciousnesses of the

physical-looking or **objective-looking, dreamal-cosmos** or **subjective-cosmos** of their **dream-sleep-state** and none else.

Similarly, from the perspective of the human consciousnesses of the **dream-sleep-state,** the only **insentient-things (**i.e the **inanimate-things** which are incapable of feeling or understanding), in existence for the purpose of interaction during their **dream-sleep-state** are the insentient, **dreamly-embodied-things** or the insentient, **subjectively-embodied-things** of the **physical-looking** or **objective-looking, dreamal-cosmos** or **subjective-cosmos** of their **dream-sleep-state,** for example, the insentient, **dremly-embodied** or the insentient, **subjectively-embodied, physical-looking** or **objective-looking** houses, roads, cars, trains, planes, mountains, rivers, moons, planets, stars, galaxies and the like of the **physical-looking** or **objective-looking, dreamal-cosmos** or **subjective-cosmos** of their **dream-sleep-state** and none else.

Hence, the unreality, reality or whatever of the two cosmoses under discussion namely, the **physical-looking** or **objective-looking, dreamal-cosmoses** or **subjective-cosmoses** of the **dream-sleep-state** on one hand and the **physical cosmos** or **objective cosmos** of mankind's **fully-awake** or **wide-awake-state** on the other, depends not at all on their **outer-look** or **external-look** or, better still, the way they **look** or **appear** to their respective observing, perceiving and experiencing human consciousnesses namely, the human consciousnesses of the **dream-sleep-state** on one hand and the human consciousnesses of the **fully-awake** or **wide-awake-state** on the other, because they both look **physical** or **objective** to their respective observing, perceiving and experiencing human consciousnesses in their respective milieus, environs, settings, locations, stamping grounds, or, if is preferred, in their respective **cosmoses.**

In the case of the **physical-looking** or **objective-looking, dreamal-cosmoses** or the **subjective-cosmoses** of the **dream-sleep-state,** the human consciousnesses of the **dream-sleep-state** firmly or unshakeably believe them to be one hundred percent **physical** and **real** or **objective** and **real** and never **dreamal** and **unreal** or **subjective** and **unreal** just as the human consciousnesses of **fully-awake** or **wide-awake-state** firmly or unshakably believe that the **cosmos** which they observe, perceive and experience in their **fully-awake** or **wide-awake-state** is one hundred percent **physical** and **real** or **objective** and **real** and never **dreamal** and **unreal** or **subjective** and **unreal.**

Furthermore, both the kinds of **cosmoses** currently under discussion, namely, the **dream-sleep-state-cosmoses** on one hand and the **fully-awake** or **wide-awake-state-cosmos** on the other, are innately, utterly transient or temporary **things** in terms of their final or ultimate, ephemerality or perishability. Thus, both these kinds of cosmoses are a match for each other, are even, equal, equivalent, comparable, matching, alike, similar, or on par, in this respect too as in all the other respects discussed so far.

It has been pointed out above that the human consciousnesses of the **dream-sleep-state** are not in the least aware or conscious of the existence of the **physical cosmos** or the **objective cosmos** of mankind's **fully-awake** or **wide-awake-state** and thus from their perspective, the **physical cosmos** or the **objective cosmos** of mankind's **fully-awake** or **wide-awake-state** is absolutely non-existent, not present, absent, without foundation, or null and void.

Therefore, from the perspective of human consciousnesses of the **dream-sleep-state**, the one and only **reality** in existence is that of the **physical-looking** or **objective-looking, dreamal-cosmos** or **subjective-cosmos** of their **dream-sleep-state** and nothing else. And they regard this **cosmos** to be one hundred percent **physical** or **objective** as well as **real** or **authentic** and not at all **dreamal** or **subjective** and **unreal** or **unauthentic** even though the truth is otherwise or, the truth is quite the opposite.

And the truth is that all the **cosmoses** of mankind's **dream-sleep-state** are mere **dreams, imageries, reveries** or **dreamries,** composed of nothing but **dream-stuff, imagery-stuff, reverie-stuff** or **dreamry-stuff** and thus, they all are in existence or, they all are present **only for a very short while** or, **transiently, temporarily** or **ephemerally,** inside the consciousnesses of these human beings themselves during their **dream-sleep-state** and nowhere else.

Therefore, all the **cosmoses** of mankind's **dream-sleep-state** are one hundred percent **dreamal** and **unreal** or **subjective** and **unauthentic** and not **physical** and **real** or **objective** and **authentic** in the manner all these human consciousnesses of the **dream-sleep-state** credit them to be, or, accept, assume, believe, guess, imagine, think, trust, or conclude them to be.

On the other hand, the human consciousnesses of the **fully-awake** or **wide-awake-state** are fully aware or conscious of the existence of the **physical-looking** or **objective-looking, dreamal-cosmoses** or **subjective-cosmoses** of mankind's **dream-sleep-state** whom they pit against or set against the **physical-cosmos** or the **objective-cosmos** of their **fully-**

awake or **wide-awake-state** in order to justify that the latter namely, the **physical-cosmos** or the **objective-cosmos** of their **fully-awake** or **wide-awake-state** is one hundred percent **physical** or **objective** and therefore **real** or **authentic** whereas the former namely, the **cosmoses** of mankind's **dream-sleep-state** are mere **physical-looking** or **objective-looking, dreamal-cosmoses** or **subjective-cosmoses** or, are mere **dreams, imageries, reveries** or **dreamries,** composed of nothing but **dream-stuff, imagery-stuff, reverie-stuff** or **dreamry-stuff** and therefore, they all are one hundred percent **dreamal** or **subjective** plus **unreal** or **unauthentic.**

The clinching or the concluding **evidence,** which human consciousnesses of the **fully-awake** or **wide-awake-state** put forward to themselves in order to justify to themselves their viewpoint that the **physical-cosmos** or the **objective-cosmos** of their **fully-awake** or **wide-awake-state** is one hundred percent **physical** or **objective** plus **real** or **authentic** whereas the **physical-looking** or **objective-looking, dreamal-cosmoses** or **subjective-cosmoses** of their **dream-sleep-state** are all, one hundred percent **dreamal** or **subjective** plus **unreal** or **unauthentic,** is that the former namely, the **cosmos** of mankind's **fully-awake** or **wide-awake-state** has already lasted some 13.7 billions light years and is likely to last many more years in the future too on account of the fact that it is not a **dream, imagery, reverie** or **dreamry** of someone else's consciousness or someone else's mind but is a **stand-alone, rock-solid, physical-reality** or **objective-reality** and therefore, a **genuine-reality** or **authentic-reality,** which can be stamped by one's physical feet and touched by one's physical fingers, not forgetting, which can also be seen by one's physical eyes, smelt by one's physical nose, tasted by one's physical tongue and heard by one's physical ears, whereas the **physical-looking** or the **objective-looking, dreamal-cosmoses** or **subjective-cosmoses** of mankind's **dream-sleep-state** last, live, endure, exist, prevail, persist, abide, or subsist only for a very short time on account of the fact that they all are mere **dreams, imageries, reveries** or **dreamries** of the human consciousnesses of **dream-sleep-state** which are observed, perceived and experienced by these human consciousnesses during their **dream-sleep-state** only and at no other time and by no one else.

However, one vital truth must always be kept in thought, celebration, awareness, or, pondering by the human consciousnesses of the **fully-awake** or **wide-awake-state.**

This vital truth is the following.

The vital truth is that the physical-looking or the objective-looking, dreamal or subjective-things of the dreamal or the subjective-cosmoses of mankind's dream-sleep-state can also be stamped by the physical-looking or the objective-looking feet & can be touched by the physical-looking or the objective-looking fingers, not to mention, can be seen by the physical-looking or the objective-looking eyes, can be smelt by the physical-looking or the objective-looking nose, can be tasted by the physical-looking or the objective-looking tongue and can be heard by the physical-looking or the objective-looking ears of the dreamal or the subjective-human-beings of the dreamal or the subjective cosmoses of mankind's dream-sleep-state in exactly the same manner as the physical or the objective-things of mankind's fully-awake or wide-awake-state can be stamped by the physical or the objective-feet and touched by the physical or the objective-fingers, not forgetting, can be seen by the physical or the objective-eyes, can be smelt by the physical or the objective-nose, can be tasted by the physical or the objective-tongue and can be heard by the physical or the objective-ears of the physical or the objective-human-beings of mankind's fully-awake or wide-awake-state.

What has been said above can be put in another way.

The **evidence,** which human consciousnesses of the **fully-awake** or **wide-awake-state** put forward, in order to justify their claim that the **physical** or the **objective cosmos** of their **fully-awake** or **wide-awake-state** is one hundred percent **real** whereas the **physical-looking** or the **objective-looking, dreamal-cosmoses** or **subjective-cosmoses** of their **dream-sleep-state,** are all one hundred percent, **unreal** or **unauthentic,** is that the former has already lasted some 13.7 billions light years and is likely to last many more billion light years in the future too on account of the fact that it is not a **dream** of someone else's consciousness or someone else's mind but is a **stand-alone, rock-solid, physical-reality** or **objective-reality** and therefore, is one hundred percent **genuine-reality** or **authentic-reality** which can be stamped by one's physical or objective feet and touched by one's physical or objective fingers, not forgetting, which can also be seen by one's physical or objective eyes, smelt by one's physical or objective nose, tasted by one's physical or objective tongue and heard by one's physical or objective ears whereas the **physical-looking** or **objective-looking, dreamal-cosmoses** or **subjective-cosmoses** of mankind's **dream-sleep-state** last only for a very short time on account of the fact that they all are mere **dreams** of human-consciousnesses of **dream-sleep-state** which are observed, perceived and experienced by the latter during their **dream-sleep-state** only and at no other time and by no one else.

However, one vital truth must always be kept in thought, cerebration, awareness, or pondering by the human consciousnesses of **fully-awake** or **wide-awake-state.**

And this vital truth is the following.

The dreamal-human-beings or the subjective-human-beings of the dreamal-cosmoses or the subjective-cosmoses of mankind's dream-sleep-state too are able to stamp on the physical-looking or the objective-looking, dreamal-things or subjective-things of these dreamal-cosmoses or subjective-cosmoses by their physical-looking or objective-looking feet, are able to touch the physical-looking or the objective-looking, dreamal-things or subjective-things of these dreamal-cosmoses or subjective-cosmoses by their physical-looking or objective-looking fingers, not forgetting, are able to see the physical-looking or the objective-looking, dreamal-things or subjective-things of these dreamal or subjective-cosmoses by their physical-looking or objective-looking eyes, are able to smell the physical-looking or the objective-looking, dreamal-things or subjective-things of these dreamal-cosmoses or subjective-cosmoses by their physical-looking or objective-looking nose, are able to taste the physical-looking or the objective-looking, dreamal-things or subjective-things of these dreamal-cosmoses or subjective-cosmoses by their physical-looking or objective-looking tongue and are able to hear the physical-looking or the objective-looking, dreamal-things or subjective-things of these dreamal-cosmoses or subjective-cosmoses by their physical-looking or objective-looking ears.

Furthermore, the human consciousnesses of the **fully-awake** or **wide-awake-state** must realise that their logic, reasoning, rationale, train of thought or thesis that all the **cosmoses** of the **dream-sleep-state** are observed, perceived and experienced by the human consciousnesses of the **dream-sleep-state** only and by no one else and that, all these **cosmoses** are observed, perceived and experienced only during **dream-sleep-state** and at no other time, is totally wide of the mark because it is based on overlooking **"the absolute** or **the most fundamental fact of the current cosmos".**

While composing and then putting forward the above kind of logic, reasoning, rationale, train of thought or thesis, the human consciousnesses of the **fully-awake** or **wide-awake-state** have overlooked **"the absolute** or **the most fundamental fact of the current cosmos"** (which is described below,) in order or with a view or desire to totally rubbish or, in order or with a view or desire to totally reject, as worthless junk, all the **dreamal** or **subjective** as well as **conscious** or **aware** plus **dreamly** or **subjectively, embodied-beings** such as human beings of the **cosmoses** of mankind's **dream-sleep-state** on one hand and all the **dreamal** or **subjective** plus **dreamly** or **subjectively, embodied** but totally **insentient** or incapable of feeling or understanding **objects** or **things** such as houses, roads, cars, trains, planes, mountains, rivers, moons, planets, stars, galaxies and the like of the **cosmoses** of mankind's **dream-sleep-state**, on the other.

Let one explain what is meant by, what has been said above.

The reason, why the above mentioned logic, ratiocination or thesis of the human consciousnesses of the **fully-awake** or **wide-awake-state** namely, or, viz. that all the **dreamal** or **subjective** as well as **conscious** or **aware** plus **dreamly** or **subjectively, embodied-beings** such as human beings of all the **cosmoses** of mankind's **dream-sleep-state** on one hand and the all the **dreamal** or **subjective** plus **dreamly** or **subjectively, embodied,** but totally **insentient** or incapable of feeling or understanding, **objects** or **things,** such as houses, roads, cars, trains, planes, mountains, rivers, moons, planets, stars, galaxies and the like of all the **cosmoses** of mankind's **dream-sleep-state** on the other, are totally rubbish or worthless junk (and therefore, not worthy of any attention) is that, this kind of logic, ratiocination or thesis of their's is based on overlooking on their part of the **"the absolute truth"** or **"the most fundamental truth"** of the current cosmos.

This **"absolute truth"** or this **"most fundamental truth"** of the current cosmos is that, at **"the absolute level"** or at **"the most fundamental level"** of the current cosmos all the **insentient** or incapable of feeling or understanding **'objects'** or **'things'** of the current cosmos are also **'the subject'** or, if one prefers, are also **"the dreamer"** or **"the spectator"** or **"the seer"** of the current cosmos. In other words, at **"the absolute level"** or at **"the most fundamental level"** of the current cosmos, all the **'objects', 'things', 'dreams', spectacles,** or **'scenes'** of the current cosmos or, all the **insentient** or incapable of feeling and understanding **'objects', 'things', 'dreams', spectacles,** or **'scenes'** of the current cosmos are also **'the subject', 'the dreamer', 'the spectator',** or **'the seer'** of the current cosmos who is none other than the **non-physical, non-material, unphysical,** or, **immaterial,** ubiquitous and infinite, **3-D** or **three-dimensional field of consciousness** or **ocean of consciousness** aka **cosmic space** which, in turn, is nothing but the **expanded, distended, dilated** or **inflated** form or version of the **dimensionless** form or version or the **original** form or version of the incredible or awe-inspiring and **the only**

one of its kind or unique **timeless, bodiless** and **dimensionless consciousness** of god, creator, maker or progenitor of the current cosmos.

What has been said above can be put in another way.

The 'supreme subject', the 'supreme dreamer', the 'supreme spectator' or the 'supreme seer' of the current cosmos "has become", all the 'objects', all the 'dreams', all the 'spectacles', or, all the 'scenes' of the current cosmos. Thus, the 'supreme subject', the 'supreme dreamer', the 'supreme spectator' or the 'supreme seer' of the current cosmos is also all the 'objects', all the 'dreams', all the 'spectacles', or, all the 'scenes' of the current cosmos or, if one prefers, is also all the insentient or incapable of feeling or understanding 'objects', 'dreams', 'spectacles' or 'scenes' of the current cosmos.

The above is the **absolute,** the **ultimate,** or, the **supreme truth.** It is not some kind of hypothetical, theoretical, abstract, academic, assumed, presumed, conjectural, notional, suppositional, untested, unproven, or, unsubstantiated speculation. Instead, it is one hundred percent, **experiential, empirical, practical, evidence-based, observed, demonstrable, hands-on** or **firsthand, absolute, ultimate** or **supreme truth.**

The above mentioned, **absolute, ultimate** or **supreme truth,** is being provided by the consciousnessbal or the awarenessbal science of **Adwait-Vedanta** to the human consciousnesses of the **fully-awake** or **wide-awake-state,**

What has been said above can be put in another way.

At the most fundamental level or **at the absolute level,** the totality of the current cosmos - consisting of the **cosmos** which is observed, perceived and experienced by the human consciousnesses in their **fully-awake** or **wide-awake-state** on one hand and all the **cosmoses** which human consciousnesses observe, perceive and experience in their **dream-sleep-states** on the other, not forgetting, all the **cosmoses** which human consciousnesses create inside their own consciousnesses during their **daydreaming activities -** is nothing but the **timeless** and **bodiless consciousness** of god, creator, maker or progenitor of the current cosmos which at the present moment exists as **non-physical, non-material, unphysical** or **immaterial,** ubiquitous and infinite, **3-D** or **three-dimensional, field of consciousness** or **ocean of consciousness** called **cosmic space** which, in turn, is nothing but the **expanded, distended, dilated** or **inflated** form or version of the **dimensionless** form or version or the **original** form or version of the incredible or awe-inspiring, **the only one of its kind** or unique, **timeless, bodiless** and **dimensionless consciousness** of stunning or amazing god, creator, maker or progenitor of the current cosmos.

In the realm or domain of **Adwait-Vedanta,** the above described, **innate** or **intrinsic** or, if it preferred, the above described **fundamental, elemental, primal, primeval,** or, **primordial** relationship that obtains between the **object** and the **subject,** the **spectacle** and the **spectator,** or, the **scene** and the **seer,** on one hand and the **subject** and the **object,** the **spectator** and the **spectacle,** or, the **seer** and the **scene** on the other, is described as the **unbreakable-relationship** or **indestructible-relationship** and is called or labelled as the **object-subject** and **subject-object** relationship or, **spectacle-spectator** and **spectator-spectacle** relationship or, **scene-seer** and **seer-scene** relationship or, **dream-dreamer** and **dreamer-dream** relationship or, **eats-eater** and

eater-eats relationship or **painting-painter** and **painter-painting** relationship and, in the end, **created-creator** and **creator-created** relationship.

The above described **relationship** is eternally or forever, innate, intrinsic, fundamental, elemental, primal, primeval, or primordial **relationship** and therefore, it is, for all time, unbreakable or indestructible.

At the level of the current cosmos, all beings, existences, or realities of the current cosmos, with the sole exception of COSMIC SPACE aka god, creator, maker or progenitor of the current cosmos are nothing but objects only vis-a-vis or in relation to COSMIC SPACE aka god, creator, maker or progenitor of the current cosmos.

The current **cosmic space** (which is nothing but god, creator, maker or progenitor of the current cosmos,) is non-physical, non-material, unphysical or immaterial, ubiquitous and infinite, 3-D or three-dimensional field of consciousness or ocean of consciousness inside which the current **physical cosmos** of mankind's **fully-awake** or **wide-awake-state,** consisting of countless moons, planets, stars, galaxies and the like, not forgetting, countless **physically** or **objectively, embodied,** conscious, aware, or, sentient beings of the current cosmos such as human beings and the like, is **floating, wafting,** or **levitating** plus **whirling, twirling** or **spiralling** non-stop as a mere **daydream, reverie, imagery, dreamry,** or **fantasy** of the current **cosmic space** aka god, creator, maker or progenitor of the current cosmos.

The current **physical cosmos** of mankind's **fully-awake** or **wide-awake-state,** consisting of countless moons, planets, stars, galaxies and the like, not forgetting, the countless

physically or **objectively, embodied,** conscious, aware, or, sentient beings of the current cosmos such as human beings and the like, is **floating, wafting,** or **levitating** plus **whirling, twirling** or **spiralling** non-stop from the beginning of the current time (which also marks the beginning of the current cosmos) as a mere **daydream, reverie, imagery, dreamry,** or **fantasy** of the current **cosmic space** aka god, creator, maker or progenitor of the current cosmos.

Thus, the **beginning** of the current **cosmos** also marks the **beginning** of the current **time**. Prior to the **beginning** of the current **cosmos,** the current **time** did not **exist.** Therefore, **time** exists only as long as the **cosmos** exists or, **time** exists merely between the **beginning** and the **ending** of the **cosmos,** not before nor after. The current **time** did not exist before the **beginning** of the current **cosmos** nor will it exist after the **ending** of the current **cosmos.** The current **time** began at the **start** of the current **cosmos** and will **end** with the **ending** of the current **cosmos.** The **life-span** of the current **time** is as much as the **life-span** of the current **cosmos,** nothing more nothing less.

Somewhere, above it has been stated that at the level of the current cosmos, all beings, existences, or realities of the current cosmos, with the sole exception of COSMIC SPACE aka god, creator, maker or progenitor of the current cosmos, are nothing but objects only vis-a-vis or in relation to COSMIC SPACE aka god, creator, maker or progenitor of the current cosmos.

Let's concentrate on that part, portion, section, or, segment of the above statement where it states that "all beings, existences, or realities of the current cosmos, are nothing but mere **'objects'** vis-a-vis or in relation to COSMIC SPACE aka god, creator, maker or progenitor of the current cosmos". In this statement, the term 'OBJECT' includes all the **physically** or **objectively, embodied,** conscious, aware, or, sentient beings of the current cosmos, for example, all the conscious, aware, or, sentient human beings of the current cosmos and not merely the INSENTIENT or incapable of feeling or understanding OBJECTS of the current cosmos such as houses, roads, cars, trains, planes, moons, planets, stars, galaxies and the like.

All the **physically** or **objectively, embodied,** conscious, aware, or, sentient human beings of the current cosmos are mere 'OBJECTS' just as are all the **physically** or **objectively, embodied,** one hundred percent INSENTIENT or one hundred percent incapable of feeling or understanding OBJECTS of the current cosmos such as houses, roads, cars, trains, planes, moons, planets, stars, galaxies and the like, vis-a-vis or in relation to COSMIC SPACE aka god, creator, maker or progenitor of the current cosmos because they all have been created, including all the **physically** or **objectively, embodied,** conscious, aware, or, sentient human beings of the current cosmos, by the COSMIC SPACE aka god, creator, maker or progenitor of the current cosmos through the instrumentality of DAYDREAMING on its part, nothing more nothing less, in order to amuse, entertain or regale itself and nothing else.

Hence, all the **physically** or **objectively, embodied,** conscious, aware, or, sentient human beings of the current cosmos are the "SUBJECTS", "SPECTATORS" or "SEERS" vis-a-vis or in relation only to all the one hundred percent "INSENTIENT" or one hundred percent incapable of feeling or understanding "OBJECTS" of the current cosmos such as houses, roads, cars, trains, planes, moons, planets, stars, galaxies and the like but they themselves are mere "OBJECTS", "SPECTACLES" or, "SCENES" vis-a-vis or in relation to COSMIC SPACE aka god, creator, maker or progenitor of the current cosmos.

From all that has been said above, it is clear that the only **evidence,** human consciousnesses of the **fully-awake** or **wide-awake-state** possess to back up their **claim** that the **cosmos** of their **fully-awake** or **wide-awake-state** is one hundred percent **physical** or **objective** and therefore, one hundred percent **real** or **authentic** and all the **cosmoses** of their **dream-sleep-state** are one hundred percent **dreamal** or **subjective** and therefore one hundred percent **unreal** or **unauthentic** is that the former namely, the **cosmos** of their **fully-awake** or **wide-awake-state** has already lasted some 13.7 billion light years and is likely to last some more light years in the future too whereas the latter namely, all the **cosmoses** of mankind's **dream-sleep-state** last only for a very short while each night.

Otherwise, as described above, in all the other respects both kinds of **cosmoses** under discussion i.e. the **cosmos** which human consciousnesses observe, perceive and experience and take part in their **fully-awake** or **wide-awake-state** on one hand and the **cosmoses** which they observe, perceive and experience and take part in their **dream-sleep-state** on the other, are one hundred percent alike.

However, the logic, ratiocination, dialectic or argumentation of human consciousnesses of **fully-awake** or **wide-awake-state** that they can decide with absolute certainty the truth vis-à-vis, regarding, concerning, with regard to, or relating to what is **physical** or **objective** and therefore one hundred percent **real** or **authentic** on one hand and what is **dreamal** or **subjective** and therefore, one hundred percent **unreal** or **unauthentic** on the other, on the basis of **time** they each last, live, exist, endure, persist, prevail, subsist or survive, betrays, displays or unintentionally reveals a great deal of intellectual short-sightedness or ignorance on the part of human consciousnesses of **fully-awake** or **wide-awake-state.**

This intellectual short-sightedness or ignorance on the part of the human consciousnesses of **fully-awake** or **wide-awake-state** with regards to their wrong logic, ratiocination, dialectic or argumentation that they can decide with absolute certainty the truth vis-à-vis, regarding, concerning, with regard to, or relating to what is **physical** or **objective** and therefore what is one hundred percent **real** or **authentic** on one hand and what is **dreamal** or **subjective** and therefore, what is one hundred percent **unreal** or **unauthentic** on the other, on the basis of **time** they each last, live, exist, endure, persist, prevail, subsist or survive is recounted below.

In regards to judging the truth of unreality, reality, or, whatever of the **cosmoses** of mankind's **dream-sleep-state** on one hand and the **cosmos** of mankind's **fully-awake** or **wide-awake-state** on the other, with absolute certainty, the total time for which these two kinds of **cosmoses** last, live, exist, endure, persist, prevail, subsist or survive is totally unimportant because in final analysis, they both are transient or temporary only and not immortal or eternal and thus are completely identical in this respect.

When human consciousnesses are in their **fully-awake** or **wide-awake-state,** what is vitally or acutely important for them to remain cognisant of, is of the fact that neither the **cosmoses** of their **dream-sleep-state** nor the **cosmos** of their **fully-awake** or **wide-awake-state** is immortal, eternal or everlasting. They both have a beginning and therefore they both have or shall have an end. Therefore, they both are real only as long as they last, no more no less.

Hence, to draw a distinction between the **cosmoses** of mankind's **dream-sleep-state** on one hand and the **cosmos** of mankind's **fully-awake** or **wide-awake-state** on the other with regards to their unreality, reality, or, whatever on the basis, foundation or fulcrum of the total time of their existence is an utter folly on the part of human consciousnesses of **fully-awake** or **wide-awake-state.**

The human consciousnesses of the **fully-awake** or **wide-awake-state** must, therefore, resist the temptation of concluding that the **cosmos,** which they observe, perceive and experience and in which they take part in their **fully-awake** or **wide-awake-state,** is one hundred percent **real** or **authentic** whereas the **cosmoses,** which they observe, perceive and experience and in which they take part in their **dream-sleep-state,** are one hundred percent **unreal** or **unauthentic** on the basis, foundation, or, fulcrum of the total time of their existence, endurance, survival or continuance.

One more point human consciousnesses of the **fully-awake** or **wide-awake-state** must take into consideration. Overlooking of this point on their part has contributed a great deal to their misapprehension that only those **cosmoses,** which they observe, perceive and experience and in which they take part during their **dream-sleep-state,** are one hundred percent **unreal** or **unauthentic** whereas the **cosmos,** which they observe, perceive and experience and in which they take part in their **fully-awake** or **wide-awake-state,** is one hundred percent **real** or **authentic.**

This point is discussed below.

It is true that the **cosmoses,** which human consciousnesses observe, perceive and experience and in which they take part in their **dream-sleep-state,** exist only for a short time each night, as measured by the time-scale or clock of the mankind's **fully-awake** or **wide-awake-state.** However, the time-scale or clock by which the human consciousnesses of the **dream-sleep-state** measure time is quite different from the time-scale or clock by which human consciousnesses of the **fully-awake** or **wide-awake-state** measure time. Hence, the events taking place inside the **cosmoses** of mankind's **dream-sleep-state,** when measured by the timescale or clock of **dream-sleep-state,** seem to human consciousnesses of **dream-sleep-state** as long, prolonged, lengthy, or, enduring as the similar events taking place in the **cosmos** of mankind's **fully-awake** or **wide-awake-state.**

Time is relative in the **current cosmos** of mankind's **fully-awake** or **wide-awake-state** just as everything else is relative in the **current cosmos** of mankind's **fully-awake** or **wide-awake-state.**

In other words, **time** is not absolute in the **current cosmos** of mankind's **fully-awake** or **wide-awake-state.** How can it be when everything else in the **current cosmos** of mankind's **fully-awake** or **wide-awake-state** is in flux, fluidity, changeability, variability, oscillation, variation, fluctuation, rise and fall, seesawing or yo-yoing.

That is to say, how can **time** be absolute in the **current cosmos** of mankind's **fully-awake** or **wide-awake-state** when everything else in the **current cosmos** of mankind's **fully-awake** or **wide-awake-state** is not absolute?

Time in the **current cosmos** of mankind's **fully-awake** or **wide-awake-state** changes, differs, fluctuates, goes up and down, swings or varies from one kind of human mood to another kind of human mood, from one state of human mind to another state of human mind, or, from one frame of human mind to another frame of human mind.

Continuing in the same vein, **time** in the **current cosmos** of mankind's **fully-awake** or **wide-awake-state** is relative, comparative, or, conditional, that is to say, it changes, fluctuates, swings or varies, from that experienced during one kind of **dream,** to that experienced during another kind of **dream** on one hand, and, from that experienced in one place to that experienced in another place, from that experienced in one planet to that experienced in another planet, from that experienced in one star to that experienced in another star, and from that experienced in one galaxy to that experienced in another galaxy inside the **very same dream,** on the other, not forgetting, from that experienced in one milieu, surroundings, environment, or, locale to that experienced in another milieu, surroundings, environment, or locale inside the **very same dream.**

To sum up.

Time is relative, comparative, or, conditional and not absolute in the **current cosmos.** The term **'current cosmos'** refers to the **cosmos** which is currently seen, observed, perceived and experienced by the human consciousnesses during their **fully-awake** or **wide-awake-state.**

Time is not absolute in the **current cosmos** of mankind's **fully-awake** or **wide-awake-state** because it cannot be otherwise. How can it be when everything else in the **current cosmos** of mankind's **fully-awake** or **wide-awake-state** is not absolute?

The concept, conception, perception, idea, impression, opinion, belief, notion, feeling, or, whim of **"passage of time"** is **quantified** or **measured** not only **objectively** or **physically** by an **objective clock** or **physical clock** in the **current cosmos** but also **quantified** or **measured, subjectively, mentally, psychologically,** or, **consciousnessbally** by a **subjective clock, mental clock, psychological clock,** or, **consciousnessbal clock** of human consciousnesses.

The **subjective clock, mental clock, psychological clock,** or, **consciousnessbal clock** of human consciousnesses is innate to all human consciousnesses and is stationed inside all human consciousnesses, irrespective of whether human consciousnesses are in their **fully-awake** or **wide-awake-state** or, in their **dream-sleep-state**, not forgetting, whether they are in their **day-dreaming-state.**

The concept, conception, perception, idea, impression, opinion, belief, notion, feeling, or, whim of **"passage of time"** is relative, comparative, or, conditional, that is to say, changes, fluctuates, swings or varies, irrespective of whether it is **quantified** or **measured** by a **physical clock** or **objective clock** or, by a **subjective clock, mental clock, psychological clock,** or, **consciousnessbal clock.**

Therefore, the concept, conception, perception, idea, impression, opinion, belief, notion, feeling, or, whim of **"passage of time"** in the **current cosmos** of mankind's **fully-awake** or **wide-awake-state** also changes, fluctuates, swings or varies from one state of human mind to another state of human mind or, from one frame of human mind to another frame of human mind, or, if it is preferred, from one state of human consciousness to another state of human consciousness or, from one frame of human consciousness to another frame of human consciousness.

The concept, conception, perception, idea, impression, opinion, belief, notion, feeling, or, whim of **"passage of time"** in the **current cosmos** of mankind's **fully-awake** or **wide-awake-state** also changes, fluctuates, swings or varies from one **daydream** of human consciousness to another **daydream** of human consciousnesses.

That is to say, the concept, conception, perception, notion, idea, impression, feeling, opinion, belief, or, whim of **"passage of time",** which human consciousnesses **quantify** or **measure** during their **daydreaming activity,** also changes, fluctuates, swings or varies from their one **daydream** to their another **daydream** and therefore, this concept, conception, perception, notion, idea, impression, feeling, opinion, belief, or, whim of **"passage of time"** is relative, comparative, or, qualified in the context of **daydream** too, and never absolute.

Likewise, the concept, conception, perception, notion, idea, impression, feeling, opinion, belief, or, whim of **"passage of time",** which human consciousnesses **quantify** or **measure** in all those **cosmoses** which human consciousnesses observe, perceive and experience in their **dream-sleep-state** is also relative, comparative or conditional.

That is to say, the concept, conception, perception, notion, idea, impression, feeling, opinion, belief, or, whim of **"passage of time"** changes, fluctuates, swings or varies from one **cosmos** of mankind's **dream-sleep-state** to another **cosmos** of mankind's **dream-sleep-state.**

The concept, conception, perception, notion, idea, impression, feeling, opinion, belief, or, whim of **"passage of time"** in the **current cosmos** of mankind's **fully-awake** or **wide-awake-state,** as **quantified** or **measured** by a **physical** or **objective clock,** has been shown to be **relative, comparative,** or, **qualified** scientifically too by the Nobel laureate Albert Einstein, the famous physicist who is the creator of the iconic equation given below :-

e = mC^2.

Hence, the pronouncement, assertion, or, claim made by mankind on the issue, topic, or, subject of **reality** or **unreality** of a particular **phenomenon, sight, spectacle, appearance,** or, **experience** on the basis, foundation, or fulcrum of the **amount** or the **quantity of time** that **phenomenon, sight, spectacle, appearance,** or, **experience** lasts, lives, exists, endures, persists, prevails, subsists, or, survives and that too by a timescale or clock which is totally different from the timescale or clock of the location, site, or, place where the **phenomenon, sight, spectacle, appearance,** or, **experience** in question, occurs, exists or takes place, is an absolute folly.

Therefore, the averment or assertion of the human consciousnesses of **fully-awake** or **wide-awake-state** that, the **cosmoses** which they observe, perceive, and experience and in which they take part during their **dream-sleep-state** are all one hundred percent **unreal** because they all are one hundred percent **dreamal-cosmoses** or **subjective-cosmoses,** composed of **dream-stuff, dreamal-stuff,** or

subjective-stuff and, as a result of which they all last, live, endure, or, exist only for a very short period of time, (namely, during the period of mankind's **dream-sleep-state** only and at no other time), whereas the **cosmos,** which they observe, perceive, & experience & in which they take part during their **fully-awake** or **wide-awake-state** is one hundred percent **real** because it is composed of one hundred percent **real** or **genuine, physical** or **objective-stuff** or **objective-matter and** as a result of which it has already lasted for a very long of time (i.e.it has already lasted some 13.7 billion light years) and is likely to last some more time, is totally false, flawed or unsound averment or assertion.

The truth is that both kinds of **cosmoses,** namely, the **cosmoses** which are observed, perceived, and experienced by the human consciousnesses during their **dream-sleep-state** on one hand, and the **cosmos** which is observed, perceived, and experienced by them during their **fully-awake** or **wide-awake-state** on the other, are on **par** with each other, or both have **parity** with each other, or both are on **equal footing** because they both**,** innately or intrinsically, are **dreamal** or **subjective, cosmoses** or, if it is preferred, because they both, innately or intrinsically, are composed of **dream, dreamal** or **subjective-stuff** and nothing but the **dream, dreamal,** or **subjective-stuff** and therefore, they both are **real** or **genuine, transiently** or **temporarily** only, and not **eternally** or **everlastingly,** meaning thereby, they both are **real** or **genuine** as long as they last, live, endure or exist and no more.

What has been said above can be put in another way.

Both kinds of **cosmoses,** namely, the **cosmoses** which are observed, perceived, and experienced by the human consciousnesses during their **dream-sleep-state** on one hand, and the **cosmos** which is observed, perceived, and experienced by them during their **fully-awake** or **wide-awake-state** on the other, are not eternally, **existent, present, observable,** or **perceivable,** or, **everlastingly, existent, present, observable,** or **perceivable** in the manner, god, creator, maker, or progenitor of both these kinds of **cosmoses** is.

And, this god, creator, maker, or progenitor of both these kinds of cosmoses, is none other than the non-physical, non-material, unphysical or immaterial, ubiquitous and infinite, 3-D or three-dimensional, field of consciousness or ocean of consciousness aka cosmic space.

The non-physical, non-material, unphysical or immaterial, ubiquitous and infinite, 3-D or three-dimensional, field of consciousness or ocean of consciousness aka cosmic space, in turn, is nothing but the expanded, distended, dilated, or inflated form or version of the dimensionless form or version or the original form or version of the incredible or awe-inspiring, the only one of its kind or unique, timeless, bodiless, and dimensionless consciousness of god, creator, maker, or progenitor of the current cosmos.

The current cosmos, is the cosmos which is observed, perceived, and experienced by the human consciousnesses in their **fully-awake** or **wide-awake-state.**

The current cosmos, namely, the **cosmos** which is observed, perceived, and experienced by the human consciousnesses in their **fully-awake** or **wide-awake-state** and in which the human consciousnesses in their **fully-awake** or **wide-awake-state,** play, participate, or engage all their lives or, better still, in which the human consciousnesses in their **fully-awake** or **wide-awake-state,** play, participate, or engage from womb to tomb, or, if it is preferred, in which the human consciousnesses in their **fully-awake** or **wide-awake-state,** eat, sleep, dream, debate and procreate, is really or genuinely a mere **daydream, reverie, imagery, dreamry** or **fantasy** of the **non-physical, non-material, unphysical,** or **immaterial,** ubiquitous and infinite, **3-D** or **three-dimensional field of consciousness** or **ocean of consciousness** aka **cosmic space** which, in turn, is nothing but the **expanded, distended, dilated,** or **inflated** form or version of the **dimensionless** form or version or the **original** form or version of the incredible or awe-inspiring, **the only one of its kind** or unique, **timeless, bodiless,** and **dimensionless consciousness** of god, creator, maker, or progenitor of the current cosmos, the current cosmos, which is seen,

observed, and experienced by the human consciousnesses in their **fully-awake** or **wide-awake-state** every moment of their lives.

THE NON-PHYSICAL TRUTHS OF THE CURRENT COSMOS

The list of all the non-physical, non-material, unphysical or immaterial truths of the current cosmos is given below:-

The non-physical, non-material, unphysical or immaterial, cosmic space of the current cosmos.

All the **non-physical, non-material, unphysical** or **immaterial,** consciousnesses of the **current cosmos,** who exist in the **current cosmos,** in the manifested, materialised, objectified, concretised, incarnate, or embodied forms, for example, all the **non-physical, non-material, unphysical** or **immaterial** human consciousnesses of the **current cosmos.**

All the **non-physical, non-material, unphysical,** or **immaterial** feelings, emotions, or sentiments (not forgetting, all the **non-physical, non-material, unphysical,** or **immaterial** thoughts, ideas, mentations, cerebrations, or reflections) of all the **non-physical, non-material, unphysical,** or **immaterial** human consciousnesses of the **current cosmos.**

All the non-physical, non-material, unphysical, or immaterial, perceptions, or impressions of the "passage of time" in the current cosmos, felt or experienced, by all the non-physical, non-material, unphysical, or immaterial human consciousnesses of the current cosmos.

THE NON-PHYSICAL, NON-MATERIAL, UNPHYSICAL, OR IMMATERIAL COSMIC SPACE OF THE CURRENT COSMOS.

The non-physical, non-material, unphysical, or immaterial cosmic space of the current cosmos is the only one of its kind, or unique, non-physical, non-material, unphysical, or immaterial truth of the current cosmos which is seen by the physical eyes of the material, objective, or concrete body of all the non-physical, non-material, unphysical, or immaterial human consciousnesses of the current cosmos.

This non-physical, non-material, unphysical, or immaterial cosmic space of the current cosmos, which is seen by the physical eyes of the material, objective, or concrete body of all the non-physical, non-material, unphysical, or immaterial human consciousnesses of the current cosmos, is ubiquitous and infinite in its expanse, spread, or range, on one hand, and 3-D or three-dimensional in its contour, construction, or configuration, on the other.

The current **3-D** or **three-dimensional,** physical, material, objective, or concrete cosmos, (consisting of countless, **3-D** or **three-dimensional,** physical, material, objective, or concrete, absolutely insentient, or incapable of feeling or understanding, moons, planets, stars, galaxies and the like on one hand and innumerable, **3-D** or **three-dimensional,** physical, material, objective, or concrete, **frames, figures, forms, bodies,** or **anatomies** of all the countless, manifested, materialised, objectified, concretised, incarnate, or embodied **consciousnesses** of the **current cosmos,** for example, all the **3-D** or **three-dimensional,** physical, material, objective, or concrete, **frames, figures, forms, bodies,** or **anatomies** of all the manifested, materialised, objectified, concretised, incarnate, or embodied, human consciousnesses, of the **current cosmos**), is floating, wafting, or levitating plus whirling, twirling, or spiralling non-stop inside this **non-physical, non-material, unphysical,** or **immaterial,** ubiquitous and infinite, **3-D** or **three-dimensional,** cosmic space of the **current cosmos** and has been doing so from the beginning of the **current time** and will continue to do so till the end of the **current time.**

ALL THE NON-PHYSICAL, NON-MATERIAL, UNPHYSICAL, OR IMMATERIAL HUMAN CONSCIOUSNESSES OF THE CURRENT COSMOS.

In total contrast to the non-physical, non-material, unphysical, or immaterial, ubiquitous and infinite, 3-D or three-dimensional, cosmic space of the current cosmos, which is seen by the physical eyes of the material, objective, or concrete body of all the non-physical, non-material, unphysical, or immaterial human consciousnesses of the current cosmos, the latter, namely, the non-physical, non-material, unphysical, or immaterial human consciousnesses of the current cosmos can never see themselves by the physical eyes of their material, objective, or concrete body.

However, even though, the non-physical, non-material, unphysical, or immaterial human consciousnesses of the current cosmos can never see themselves by the physical eyes of their material, objective, or concrete body, they nevertheless are aware, conscious or cognisant of their own existence, presence, isness, or, beingness as the aware, conscious or cognisant, beings, truths, or realities because they innately are self-aware-beings, self-aware-truths or self-aware-realities.

Hence, unlike the physical matter/physical energy duo of the current cosmos (both of which are innately insentient or incapable of feeling or understanding beings, things, truths, or realities), all the non-physical, non-material, unphysical, or immaterial human consciousnesses of the current cosmos are self-illumined-beings or self-illumined-truths, who know the truth-of-their-own-existence, the truth-of-their-own-presence, the truth-of-their-own-is-ness, or the truth-of-their-own-being-ness with the help of their own innate-power, force, might, or ability called the power, force, might, or ability of self-awareness or self-cognisance.

What has been said above can be put in another way.

Hence, unlike the physical matter/physical energy duo of the current cosmos, (both of which are innately insentient or incapable of feeling or understanding beings, things, truths, or realities), all the non-physical, non-material, unphysical, or immaterial human consciousnesses of the current cosmos are self-illumined-beings or self-illumined-truths who know the truth-of-their-own-existence, the truth-of-their-own-presence, the truth-of-their-own-is-ness, or the truth-of-their-own-being-ness with the help of their own innate-light, innate-luminance, or innate-luminosity called the light, luminance, or luminosity of self-awareness or self-cognisance.

In the realm or domain of Adwait-Vedanta, all the non-physical, non-material, unphysical, or immaterial human consciousnesses of the current cosmos are called Swayam-Prakasat or Swayam-Prakasit beings or truths (i.e. Swayam+Prakas+Sat), Swayam-Bhusat beings or truths (i.e. Swayam+Bhu+Sat), or, Swayam-Chitsat beings or truths (i.e. Swayam+Chit+Sat) which means that all the non-physical, non-material, unphysical or immaterial human consciousnesses of the current cosmos are self-illumined, self-knowing, self-cognisant, or self-aware-beings, or, if it is preferred, all the non-physical, non-material, unphysical or immaterial human consciousnesses of the current cosmos are self-illumined, self-knowing, self-cognisant, or self-aware-truths.

All the **non-physical, non-material, unphysical,** or **immaterial,** human consciousnesses of the **current cosmos** are identical or akin to, or, if it is preferred, are carbon copy or splitting image of their **source** or **fountainhead,** namely, the **non-physical, non-material, unphysical,** or **immaterial,** ubiquitous and infinite, **3-D** or **three-dimensional** cosmic space of the **current cosmos,** but purely, merely, or only, with regards to three aspects of cosmic space, or, if it is preferred, purely, merely, or only, in the matters of, or vis-a-vis three aspects of cosmic space i.e. firstly, with regards to being **self-aware** or **self-cognisant** of their own existence, presence, is-ness, or being-ness, secondly, with regards to being timeless, deathless, immortal or eternal, and finally, with regards to having, or possessing feelings, emotions, or sentiments.

The implication, significance, or import of what has been said immediately above, and one quotes here in verbatim :- "All the **non-physical, non-material, unphysical,** or **immaterial,** human consciousnesses of the **current cosmos** are identical or akin to, or, if it is preferred, are carbon copy or splitting image of their **source** or **fountainhead,** namely, the **non-physical, non-material, unphysical,** or **immaterial,** ubiquitous and infinite, **3-D** or **three-dimensional,** cosmic space of the **current cosmos,** but purely, merely, or only, with regards to three aspects of cosmic space, or, if it is preferred, purely, merely, or only, in the matters of, or vis-a-vis three aspects of cosmic space i.e. firstly, with regards to being **self-aware** or **self-cognisant** of their own existence, presence, is-ness, or being-ness, secondly, with regards to being timeless, deathless, immortal or eternal, and finally, with regards to having or possessing feelings, emotions, or sentiments", is that the **non-physical, non-material, unphysical,** or **immaterial,** ubiquitous and infinite, **3-D** or **three-dimensional** cosmic space of the **current cosmos** is not only a **timeless, deathless, immortal** or **eternal being,** but it also is an incredible or awe-inspiring, **self-illumined-being,** or **self-illumined-truth,** who knows the **truth-of-its-own-existence,** the **truth-of-its-own-presence,** the **truth-of-its-own-isness,** or, the **truth-of-its-own-beingness** by its own **innate-light** or, by its own **innate-power** called the **innate-light** or **innate-power** of **self-awareness** or, the **innate-light** or **innate-power** of **self-cognisance,** despite the fact that it is a soundless, silent, mum, or mute **being** or **truth,** due to the lack on its part of physical, material, objective, or concrete **speech-equipment** i.e. lips and lungs plus larynx and tongue. Not only that, **cosmic space** also possesses intelligence plus feelings, emotions, or sentiments in infinite measure.

Therefore, the non-physical, non-material, unphysical, or immaterial, ubiquitous and infinite, 3-D or three-dimensional, cosmic space of the current cosmos is not an insentient, inanimate, or incapable of understanding or feeling being, thing, entity, reality, or truth as believed or assumed by the human consciousnesses of the current cosmos.

Instead, the non-physical, non-material, unphysical, or immaterial, ubiquitous and infinite, 3-D or three-dimensional, being, thing, entity, reality, or truth of the current cosmos called cosmic space is an incredible or awe-inspiring, expanded, distended, dilated, or inflated form or version of the dimensionless form or version, or the original form or version of the timeless and bodiless, consciousness of god, creator, maker, or progenitor of the current cosmos. The dimensionless form or version, or, the original form or version of the timeless and bodiless, consciousness of god, creator, maker, or progenitor of the current cosmos was in existence in the past only i.e. some 13.7 billion light years ago. It does not exist now.

To repeat.

The **dimensionless** form or version, or, the **original** form or version of the **timeless** and **bodiless consciousness** of god, creator, maker, or progenitor of the **current cosmos** does not exist now. This form or version i.e. the **dimensionless** form or version, or the **original** form or version of the **timeless** and **bodiless consciousness** of god, creator, maker, or progenitor of the **current cosmos** existed some 13.7 billion light years ago only. It does not exist now. The incredible or awe-inspiring and **the only one of its kind** or unique, **timeless** and **bodiless consciousness** of god, creator, maker, or progenitor of the **current cosmos** exists presently, currently, or now in its **3-D** or **three-dimensional** form or version called **cosmic space** which is seen by the **physical eyes** of the **material, objective,** or **concrete body** of all the **non-physical, non-material, unphysical,** or **immaterial** human consciousnesses of the **current cosmos.**

The physical eyes of the material, objective, or concrete body of all the non-physical, non-material, unphysical, or immaterial human consciousnesses of the current cosmos are able to see cosmic space or are able to see the 3-D or three-dimensional form or version of the timeless and bodiless consciousness of god, creator, maker, or progenitor of the current cosmos, because the 3-D or three-dimensional physical eyes of the 3-D or three-dimensional, material, objective, or concrete bodies of all the non-physical, non-material, unphysical, or immaterial human consciousnesses of the current cosmos are housed inside this cosmic space or inside this 3-D or three-dimensional form or version of the timeless and bodiless consciousness of god, creator, maker, or progenitor of the current cosmos.

Thus, the current **cosmic space** is the **timeless** and **bodiless consciousness** of god, creator, maker, or progenitor of the **current cosmos,** in its **3-D** and **three-dimensional** form or version, nothing more nothing less.

The **timeless** and **bodiless consciousness** of god, creator, maker, or progenitor of the **current cosmos** has created its current **3-D** or **three-dimensional** form or version, out of its **dimensionless** form or version or, out of its **original** form or version, through the process of **expansion, distension, dilation,** or **inflation** of its **dimensionless** form or version or its **original** form or version.

As said before, the **original** form or version which was the **dimensionless** form or version of the **timeless** and **bodiless consciousness** of god, creator, maker, or progenitor of the **current cosmos** existed some 13.7 billion light years ago only and does not exist now.

Irrespective of whether the incredible or awe-inspiring, and **the only one of its kind** or unique, **timeless** and **bodiless consciousness** of god, creator, maker, or progenitor of the current cosmos exists in its **dimensionless** form or version i.e. the **original** form or version, or, is in its **current** form or version i.e. the **3-D** or **three-dimensional** form or version aka current **cosmic space** of mankind's **fully-awake-awake** or **wide-awake-state,** it is an extraordinary or amazing and the only one of its kind or unique, **timeless** and **bodiless consciousness** of infinite intelligence and emotions, feelings, or sentiments.

What has been said above can be put in another way.

The current **cosmic space** of mankind's **fully-awake** or **wide-awake-state,** is a soundless, silent, mum, or mute, **being, truth, reality,** or **existence,** due to the lack on its part of a physical, material, objective, or concrete, **speech-equipment,** namely, larynx and tongue plus lips and lungs. It lacks the physical, material, objective, or concrete, **speech-equipment,** namely, larynx and tongue plus lips and lungs, because, unlike the human consciousnesses of the current cosmos, it is an incredible, or awe-inspiring, and **the only one of its kind** or unique, **timeless** and **bodiless** or **unembodied, disembodied, unbodied,** or **disbodied consciousness** of infinite intelligence and emotions, feelings, or sentiments. There is no one like **him, her, it,** or **whatever** in the current cosmos. That is why it is titled or tagged as '**the only one of its kind**' or '**unique**'. However, one cannot say whether there is someone like **him, her, it,** or **whatever** beyond the current cosmos.

ALL THE **NON-PHYSICAL, NON-MATERIAL, UNPHYSICAL, OR IMMATERIAL,** FEELINGS, EMOTIONS OR SENTIMENTS (not forgetting, all the **non-physical, non-material, unphysical,** or **immaterial,** thoughts, ideas, mentations, cerebrations, or reflections) OF ALL THE **NON-PHYSICAL, NON-MATERIAL, UNPHYSICAL, OR IMMATERIAL** HUMAN CONSCIOUSNESSES OF THE **CURRENT COSMOS,** ON ONE HAND, and ALL THE **NON-PHYSICAL, NON-MATERIAL, UNPHYSICAL, OR IMMATERIAL** PERCEPTIONS OR IMPRESSIONS OF THE **"PASSAGE OF TIME"** IN THE CURRENT COSMOS, FELT OR EXPERIENCED BY ALL THE **NON-PHYSICAL, NON-MATERIAL, UNPHYSICAL, OR IMMATERIAL** HUMAN CONSCIOUSNESSES OF THE **CURRENT COSMOS**, ON THE OTHER.

All the **non-physical, non-material, unphysical** or **immaterial,** feelings, emotions, or sentiments (not forgetting, all the **non-physical, non-material, unphysical,** or **immaterial** thoughts, ideas, mentations, cerebrations, or reflections) of all the **non-physical, non-material, unphysical** or **immaterial,** human consciousnesses of the **current cosmos,** on one hand, and all the **non-physical, non-material, unphysical,** or **immaterial** perceptions, or impressions of the **"passage of time"** in the **current cosmos,** felt or experienced, by all the **non-physical, non-material, unphysical** or **immaterial,** human consciousnesses of the **current cosmos**, on the other, are **merely** understood, comprehended, or sensed by the latter, namely, the **non-physical, non-material, unphysical,** or **immaterial** human consciousnesses of the **current cosmos,** and never seen by them through the **physical eyes** of their **material, objective,** or **concrete body.**

What has been said above can be expressed in another way.

None of the **non-physical, non-material, unphysical** or **immaterial** feelings, emotions, or sentiments (not forgetting, all the **non-physical, non-material, unphysical,** or **immaterial** thoughts, ideas, mentations, cerebrations, or reflections) of the **non-physical, non-material, unphysical** or **immaterial** human consciousnesses of the **current cosmos,** on one hand, and none of the **non-physical, non-material, unphysical** or **immaterial** perceptions, or impressions of the **"passage of time"** in the current cosmos, felt or experienced, by the **non-physical, non-material, unphysical,** or **immaterial** human consciousnesses of the **current cosmos,** on the other, are ever seen by the latter, namely, the **non-physical, non-material, unphysical** or **immaterial** human consciousnesses of the **current cosmos,** through the **physical eyes** of their **material, objective,** or **concrete body** in the manner, the **non-physical, non-material, unphysical** or **immaterial,** ubiquitous and infinite, **3-D** or **three-dimensional,** cosmic space of the **current cosmos** is seen by them through the **physical eyes** of their **material, objective,** or **concrete body.**

In other words, all the **non-physical, non-material, unphysical** or **immaterial,** human consciousnesses of the **current cosmos,** on one hand, and all the **non-physical, non-material, unphysical** or **immaterial,** feelings, emotions, or sentiments (not forgetting, all the **non-physical, non-material, unphysical,** or **immaterial** thoughts, ideas, mentations, cerebrations, or reflections) of these **non-physical, non-material, unphysical** or **immaterial,** human consciousnesses of the **current cosmos** on the other hand, not forgetting, all the **non-physical, non-material, unphysical** or **immaterial**, perceptions, or impressions of the **"passage of time"** in the current cosmos, felt or experienced, by these **non-physical, non-material, unphysical,** or **immaterial,** human consciousnesses of the **current cosmos**, can never be seen by the **physical eyes** of the **material, objective,** or **concretised body** of these **non-physical, non-material, unphysical** or **immaterial** human consciousnesses of the **current cosmos,** in the manner, the **non-physical, non-material, unphysical** or **immaterial,** ubiquitous and infinite, **3-D** or **three-dimensional,** cosmic space of the **current cosmos** is seen by the **physical eyes** of the **material, objective,** or **concrete body** of these **non-physical, non-material, unphysical,** or **immaterial** human consciousnesses of the **current cosmos.**

WHAT IS THE IMPORT OR SIGNIFICANCE OF THE ABOVE -MENTIONED DICHOTOMY?

The most important question before the **non-physical, non-material, unphysical** or **immaterial** human consciousnesses of the **current cosmos** is that :-

What is the import or significance of the above mentioned dichotomy that the non-physical, non-material, unphysical or immaterial, ubiquitous and infinite, 3-D or three-dimensional, cosmic space of the current cosmos, is the only one of its kind or unique, non-physical, non-material, unphysical or immaterial being, truth, or reality of the current cosmos which is seen by the physical eyes of the material, objective, or concrete body of all the non-physical, non-material, unphysical or immaterial, human consciousnesses of the current cosmos, whereas, in mystifying, baffling, puzzling, or confounding contrast or comparison, all the non-physical, non-material, unphysical or immaterial human consciousnesses of the current cosmos can never see themselves by the physical eyes of their material, objective, or concrete body.

Not only that, the non-physical, non-material, unphysical or immaterial, human consciousnesses of the current cosmos also can never see by the physical eyes of their material, objective, or concrete body, all their non-physical, non-material, unphysical or immaterial feelings, sentiments or emotions (not forgetting, all their non-physical, non-material, unphysical, or immaterial thoughts, ideas, mentations, cerebration, or reflections), on one hand, and all their non-physical, non-material, unphysical or immaterial, perceptions, or impressions of the "passage of time" in the current cosmos, felt or experienced, by them, on the other.

What has been said above can be put in another way.

What is the import or significance of the above mentioned dichotomy that the non-physical, non-material, unphysical or immaterial, ubiquitous and infinite, 3-D or three-dimensional, cosmic space of the current cosmos, is the only one of its kind or unique, non-physical, non-material, unphysical or immaterial being, truth, or reality of the current cosmos which is seen by the physical eyes of the material, objective, or concrete body of all the non-physical, non-material, unphysical or immaterial, human consciousnesses of the current cosmos, whereas, in mystifying, baffling, puzzling, or confounding contrast or comparison, all the non-physical, non-material, unphysical or immaterial human consciousnesses of the current cosmos can never see themselves with the aid of the physical eyes of their material, objective, or concrete body, notwithstanding the fact that, they all are aware, conscious or cognisant beings, truths, or realities, who are aware, conscious or cognisant of the truth, reality, or fact of their own existence, presence, is-ness, or, being-ness as the aware, conscious or cognisant, beings, truths, or realities, on account of the fact that they all are innately, self-aware-beings, self-aware-truths, or self-aware-realities.

Furthermore, all the **non-physical, non-material, unphysical** or **immaterial,** feelings, sentiments or emotions (not forgetting, all the **non-physical, non-material, unphysical,** or **immaterial** thoughts, ideas, mentations, cerebrations, or reflections), of the **non-physical, non-material, unphysical** or **immaterial** human consciousnesses of the **current cosmos** on one hand, and all the **non-physical, non-material, unphysical** or **immaterial,** perceptions, or impressions of the **"passage of time"** in the current cosmos, felt or experienced, by all the **non-physical, non-material, unphysical,** or **immaterial** human consciousnesses of the **current cosmos**, on the other, are **merely** understood, comprehended, or sensed by all the **non-physical, non-material, unphysical** or **immaterial** human consciousnesses of the **current cosmos,** but never seen by them, that is to say, but never seen by the **non-physical, non-material, unphysical,** or **immaterial** human consciousnesses of the **current cosmos** with the aid of the **physical eyes** of their **material, objective,** or **concrete body,** in the manner, the **non-physical, non-material, unphysical** or **immaterial,** ubiquitous and infinite, **3-D** or **three-dimensional,** cosmic space of the **current cosmos** is seen by the **non-physical, non-material, unphysical,** or

immaterial human consciousnesses of the **current cosmos** with the aid of the **physical eyes** of their **material, objective,** or **concrete body.**

Therefore, as said before, the most important question facing all the **non-physical, non-material, unphysical** or **immaterial** human consciousnesses of the **current cosmos** is :-

"What is the import or significance of the above mentioned dichotomy that the non-physical, non-material, unphysical or immaterial, ubiquitous and infinite, 3-D or three-dimensional, cosmic space of the current cosmos, is the only one of its kind or unique, non-physical, non-material, unphysical or immaterial being, truth, or reality of the current cosmos which is seen by the physical eyes of the material, objective, or concrete body of all the non-physical, non-material, unphysical or immaterial, human consciousnesses of the current cosmos, whereas, in mystifying, baffling, puzzling, or confounding contrast or comparison, all the non-physical, non-material, unphysical or immaterial human consciousnesses of the current cosmos can never see themselves by the physical eyes of their material, objective, or concrete body"?

In other words, why is it that the **non-physical, non-material, unphysical** or **immaterial,** human consciousnesses of the **current cosmos** are not able to see themselves by the **physical eyes** of their **material, objective,** or **concrete body** in the manner they are able to see the **non-physical, non-material, unphysical** or **immaterial,** ubiquitous and infinite, **3-D** or **three-dimensional** cosmic space by the **physical eyes** of their **material, objective,** or **concrete body.**

And, furthermore, what is the import or significance of the dichotomy, that the non-physical, non-material, unphysical or immaterial, ubiquitous and infinite, 3-D or three-dimensional, cosmic space of the current cosmos, is the only one of its kind or unique, non-physical, non-material, unphysical or immaterial being, truth, or reality of the current cosmos, which is seen by the physical eyes of the material, objective, or concrete body of all the non-physical, non-material, unphysical or immaterial, human consciousnesses of the current cosmos, whereas, in mystifying, baffling, puzzling, or confounding contrast or comparison, all the non-physical, non-material, unphysical or immaterial feelings, sentiments or emotions (not forgetting, all the non-physical, non-material, unphysical, or immaterial thoughts, ideas, mentations, cerebrations, or reflections), of the non-physical, non-material, unphysical or immaterial human consciousnesses of the current cosmos, on one hand, and all the non-physical, non-material, unphysical or immaterial, perceptions, or impressions of the "passage of time" in the current cosmos, felt or experienced, by all the non-physical, non-material, unphysical, or immaterial human consciousnesses of the current cosmos, on the other, are merely understood, comprehended, or sensed by all the non-physical, non-material, unphysical or immaterial human consciousnesses of the current cosmos, but never seen by the physical eyes of their material, objective, or concrete body in the manner, the non-physical, non-material, unphysical or immaterial, ubiquitous & infinite, 3-D or three-dimensional, cosmic space of the current cosmos is seen by the physical eyes of their material, objective, or concrete body.

All the **non-physical, non-material, unphysical** or **immaterial** feelings, sentiments or emotions (not forgetting, all the **non-physical, non-material, unphysical,** or

immaterial thoughts, ideas, mentations, cerebrations, or reflections), of all the **non-physical, non-material, unphysical** or **immaterial** human consciousnesses of the **current cosmos** on one hand, and all the **non-physical, non-material, unphysical** or **immaterial,** perceptions, or impressions of the **"passage of time"** in the current cosmos, felt or experienced, by all the **non-physical, non-material, unphysical,** or **immaterial** human consciousnesses of the **current cosmos,** on the other, can never be seen by the **physical eyes** of the **material, objective,** or **concrete body** of all the **non-physical, non-material, unphysical** or **immaterial** human consciousnesses of the **current cosmos,** in the manner, the **non-physical, non-material, unphysical** or **immaterial,** ubiquitous and infinite, **3-D** or **three-dimensional,** cosmic space of the **current cosmos** is seen by the **physical eyes** of the **material, objective,** or **concrete body** of all the **non-physical, non-material, unphysical** or **immaterial** human consciousnesses of the **current cosmos.**

So the question is :-

"Why this dichotomy"?
"What is its import or significance"?

The Answer is :-

God, creator, maker, or progenitor of the current cosmos, namely, the non-physical, non-material, unphysical or immaterial, ubiquitous and infinite, 3-D or three-dimensional field of consciousness or ocean of consciousness aka current cosmic space of mankind's fully-awake or wide-awake-state has created the above mentioned dichotomy in the current cosmos on purpose, by design, deliberately, calculatingly, or wittingly, so that the non-physical, non-material, unphysical or immaterial human consciousnesses of the current cosmos will never have to search the creator, maker, or progenitor of the current cosmos anywhere i.e. so that the creator, maker, or progenitor of the current cosmos will always be accessible to them first-hand, face-to-face, without a go-between, or, without an intermediary, intercessor, mediator, negotiator, arbitrator, broker, medium, minister, messenger, proxy, or, salesperson, every moment of their lives in the current cosmos and also, so that, they will be able to see him, her, it, or, whatever by the physical eyes of their material, objective, or concrete body, every moment of their lives.

What has been said above can be put in another way.

God, creator, maker, or progenitor of the current cosmos, namely, the non-physical, non-material, unphysical or immaterial, ubiquitous and infinite, 3-D or three-dimensional field of consciousness or ocean of consciousness aka the current cosmic space of mankind's fully-awake or wide-awake-state has given birth to the above mentioned dichotomy or duality on purpose, by design, deliberately, calculatingly, or wittingly, or, if it is preferred, has given birth to the above mentioned paradox, enigma, mystery, or conundrum on purpose, by design, deliberately, calculatingly, or wittingly, so that, it, namely, the cosmic space or god, creator, maker, or progenitor of the current cosmos will always be accessible or available to each and every human consciousness of the current cosmos every moment of its existence, presence, is-ness, or being-ness.

In other words, god, creator, maker, or progenitor of the current cosmos, namely, the non-physical, non-material, unphysical or immaterial, ubiquitous and infinite, 3-D or three-dimensional field of consciousness or ocean of consciousness aka the current cosmic space of mankind's fully-awake or wide-awake-state has created the above mentioned dichotomy or duality, or, if one prefers, paradox, enigma, mystery, or conundrum in the current cosmos, on purpose, by design, deliberately, calculatingly, or wittingly, so that, the non-physical, non-material, unphysical or immaterial human consciousnesses of the current cosmos will never have to search anywhere their own source or fountainhead, on one hand, and will also never have to search anywhere the creator, maker, or progenitor of their physical, material, concrete, or objective bodies, on the other, not forgetting, the creator, maker, or progenitor of the physical, material, concrete, or objective bodies of all the other manifested, materialised, concretised, objectified, incarnate, or embodied consciousnesses of the current cosmos, for example, all the animal consciousnesses of the current cosmos, and the creator, maker, or progenitor of all the absolutely insentient or incapable of feeling or understanding, physical, material, concrete, or objective items or things of the current cosmos, for example, all the moons, planets, stars, galaxies and the like of the current cosmos.

That is to say, god, creator, maker, or progenitor of the current cosmos, namely, the non-physical, non-material, unphysical or immaterial, ubiquitous and infinite, 3-D or three-dimensional field of consciousness or ocean of consciousness aka the current cosmic space of mankind's fully-awake or wide-awake-state has created the above mentioned dichotomy or duality, or, if one prefers, paradox, enigma, mystery, or conundrum in the current cosmos, on purpose, by design, deliberately, calculatingly, or wittingly, so that, to the non-physical, non-material, unphysical or immaterial human consciousnesses of the current cosmos, their own source or fountainhead will always be accessible or available to them every moment of their existence, presence, is-ness, or being-ness in the current cosmos.

Not only that, god, creator, maker, or progenitor of the current cosmos, namely, the non-physical, non-material, unphysical or immaterial, ubiquitous and infinite, 3-D or three-dimensional field of consciousness or ocean of consciousness aka the current cosmic space of mankind's fully-awake or wide-awake-state has created the above mentioned dichotomy or duality, or, if one prefers, paradox, enigma, mystery, or conundrum in the current cosmos, on purpose, by design, deliberately, calculatingly, or wittingly, so that, it itself, will always be accessible or available to all the non-physical, non-material, unphysical or immaterial human consciousnesses of the current cosmos every moment of their existence, presence, is-ness, or being-ness in the current cosmos.

The seeing, witnessing, noticing, or viewing of the non-physical, non-material, unphysical or immaterial, ubiquitous and infinite, 3-D or three-dimensional field of consciousness or ocean of consciousness aka current cosmic space of mankind's fully-awake or wide-awake-state, by the physical eyes of the material, objective, or concrete body of all the non-physical, non-material, unphysical or immaterial human consciousnesses of the current cosmos, is essential, vital, or critical, for all the non-physical, non-material, unphysical or immaterial human consciousnesses of the current cosmos, on account of the fact that, the non-physical, non-material, unphysical or immaterial, ubiquitous and infinite, 3-D or three-dimensional, field of consciousness or ocean of consciousness aka current cosmic space of mankind's fully-awake or wide-awake-state is, not only the source or the fountainhead of all the individual, separate, discrete, or independent, non-physical, non-material, unphysical or immaterial, manifested, materialised, concretised, objectified, incarnate, or, embodied consciousnesses of the current cosmos, for example, all the individual, separate, discrete, or independent, non-physical, non-material, unphysical or immaterial, manifested, materialised, concretised, objectified, incarnate, or, embodied human consciousnesses of the current cosmos, but it also is the god, creator, maker, or progenitor of all the physical, material, concrete, or objective bodies of all the non-physical, non-material, unphysical or immaterial, manifested, materialised, concretised, objectified, incarnate, or, embodied consciousnesses of the current cosmos, for example, the physical, material, concrete, or objective bodies of all the manifested, materialised, concretised, objectified, incarnate, or, embodied human consciousnesses of the current cosmos, not forgetting, it also is god, creator, maker, or progenitor of all the physical, material, concrete, or objective bodies of all the other non-physical, non-material, unphysical or immaterial, manifested, materialised, concretised, objectified, incarnate, or embodied consciousnesses of the current cosmos, for example, all the animal consciousnesses of the current cosmos, and all the absolutely insentient or incapable of feeling or understanding, physical, material, concrete, or objective items or things of the current cosmos, for example, all the moons, planets, stars, galaxies and the like of the current cosmos.

Therefore, the non-physical, non-material, unphysical or immaterial, plus, the timeless and bodiless, or, if one prefers, plus the timeless and disembodied, disbodied, unembodied, unbodied, or discarnate consciousness of god, creator, maker, or progenitor of the current cosmos, namely, the non-physical, non-material, unphysical or immaterial, ubiquitous and infinite, 3-D or three-dimensional field of consciousness or ocean of consciousness aka the current cosmic space of mankind's fully-awake or wide-awake-state, has willed that, each and every human consciousness of the current cosmos will have free access to him, her, it, or, whatever, every moment of its existence, presence, is-ness, or being-ness in the current cosmos.

What has been said above can be put in another way.

Furthermore, god, creator, maker, or progenitor of the current cosmos, namely, the non-physical, non-material, unphysical or immaterial, ubiquitous and infinite, 3-D or three-dimensional field of consciousness or ocean of consciousness aka current cosmic space of mankind's fully-awake or wide-awake-state has created the above mentioned dichotomy or duality, or, if one prefers, paradox, enigma, mystery, or conundrum in the current cosmos, on purpose, by design, deliberately, calculatingly, or wittingly, so that, to the non-physical, non-material, unphysical or immaterial human consciousnesses of the current cosmos, the source or the fountainhead of all the individual, separate, discrete, or independent, consciousnesses of the current cosmos, or, better still, the source or the fountainhead of all the manifested, materialised, concretised, objectified, incarnate, or, embodied consciousnesses of the current cosmos, for example, all the human consciousnesses of the current cosmos on one hand, and god, creator, maker, or progenitor of their physical, material, concrete, or objective bodies, not forgetting, god, creator, maker, or progenitor of the physical, material, concrete, or objective bodies of all the other manifested, materialised, concretised, objectified, incarnate, or embodied consciousnesses of the current cosmos, for example, all the animal consciousnesses of the current cosmos, and god, creator, maker, or progenitor of all the absolutely insentient or incapable of feeling and understanding, physical, material, concrete, or objective items or things of the current cosmos, for example, all the moons, planets, stars, galaxies and the like, of the current cosmos, on the other, will always be accessible or available to all the human consciousnesses of the current cosmos freely, without obstruction or impediment, or, without any let or hindrance, every moment of their existence, presence, is-ness, or being-ness in the current cosmos.

Just to remind, all the human consciousnesses of the **current cosmos** that this **source** or the **fountainhead** of theirs, on one hand, and the **source** or the **fountainhead** of innumerable other individual, separate, discrete, or independent, consciousnesses of the **current cosmos,** for example, all the animal consciousnesses of the **current cosmos,** on the other, not forgetting, god, creator, maker, or progenitor, or, if it is preferred, the **Supreme Being** behind the creation of the remainder of the **current cosmos,** that is to say, the **Supreme Being** behind the creation of the **physical, material, concrete,** or, **objective aspect** of the **current cosmos,** is none other than the **non-physical, non-material, unphysical** or **immaterial,** ubiquitous and infinite, **3-D** or **three-dimensional field of consciousness** or **ocean of consciousness** aka the current **cosmic space** of mankind's **fully-awake** or **wide-awake-state,** which is seen by all the human consciousnesses of the **current cosmos** in their **fully-awake** or **wide-awake-state,** with the aid of the **physical eyes** of their **material, objective,** or **concrete body.** This incredible **cosmic space,** is nothing but the **expanded, distended, dilated,** or **inflated** form or version of the **dimensionless** form or version or the **original** form or version of the awe-inspiring and **the only one of its kind** or unique, **timeless, bodiless,** or **dimensionless**

consciousness of the **Supreme Being**, god, creator, maker, or progenitor of the **current cosmos.**

'O', 'COSMIC SPACE ! I LOVE YOU. I AM YOUR DREAM AND YOU ARE MY DREAMER'

From antiquity, the **non-physical, non-material, unphysical** or **immaterial** human consciousnesses of the **current cosmos** have addressed the creator, maker, or progenitor of the **current cosmos** by countless names and have offered their prayer to this creator, maker, or progenitor of the **current cosmos** in countless ways as per their liking, fondness, or fancy or, if it is preferred, as per their custom, belief, or faith.

In the domain or realm of the consciousnessbal or awarenessbal science of **Adwait-Vedanta,** the above-mentioned prayer, namely, **'O', 'Cosmic Space! I love you. I am your dream and you are my dreamer',** offered to the creator, maker, or progenitor of the **current cosmos** is the preferred choice.

The creator, maker, or progenitor of the current cosmos is the conscious, aware, or sentient plus non-physical, non-material, unphysical or immaterial, ubiquitous and infinite, 3-D or three-dimensional field of consciousness or ocean of consciousness aka the current cosmic space of mankind's fully-awake or wide-awake-state.

Therefore, the above prayer is offered to the conscious, aware, or sentient plus non-physical, non-material, unphysical or immaterial, ubiquitous and infinite, 3-D or three-dimensional field of consciousness or ocean of consciousness aka the current cosmic space of mankind's fully-awake or wide-awake-state by the pathfinders or the pioneers of the consciousnessbal or the awarenessbal science of Adwait-Vedanta, who have consciousnessbally or awarenessbally explored and then consciousnessbally or awarenessbally experimented in the field of the consciousnessbal or the awarenessbal science of Adwait-Vedanta for centuries and are still doing so and one has no doubt, will continue to do so in the future too.

The conscious, aware, or sentient plus non-physical, non-material, unphysical

I or immaterial, ubiquitous and infinite, 3-D or three-dimensional field of consciousness or ocean of consciousness aka the current cosmic space of mankind's fully-awake or wide-awake-state is the expanded, distended, dilated, or inflated form or version of the dimensionless form or version or the original form or version of the incredible or awe-inspiring, and the only one of its kind or unique, timeless, bodiless, and dimensionless consciousness of god, creator, maker, or progenitor of the current cosmos.

The **current cosmos,** is the **cosmos** which is seen, observed, and experienced by the current human consciousnesses in their **fully-awake** or **wide-awake-state,** every moment of their lives.

The above Adwait-Vedantic prayer i.e. 'O', 'Cosmic Space! I love you. I am your dream and you are my dreamer', offered to the conscious, aware, or sentient, creator, maker, or progenitor of the current cosmos, namely, the conscious, aware, or sentient plus the non-physical, non-material, unphysical or immaterial, ubiquitous and infinite, 3-D or three-dimensional field of consciousness or ocean of consciousness aka the current cosmic space of mankind's fully-awake or wide-awake-state, anchors, embeds, plants, or inserts, or, if it is preferred, hammers in, into the genius, animus, élan vital, noumenon, or heart of the non-physical, non-material, unphysical or immaterial human consciousnesses of the current cosmos, the basic, elemental, fundamental, or vital plus experiential, empirical, experimental, observational, posteriori, or, the sense-perception-based-fact, the sense-perception-based-truth, or the sense-perception-based-reality, or, if it is preferred, the sense-perception-based-evidence, the sense-perception-based-proof, or the sense-perception-based-authentication or, better still, the sense-perception-based-knowledge, the sense-perception-based-insight, or the sense-perception-based-know-how, regarding or vis-a-vis, the creator, maker, or, progenitor of the current cosmos in the simplest possible way.

This basic, elemental, fundamental, or vital plus experiential, empirical, experimental, observational, posteriori, or, the sense-perception-based-fact, the sense-perception-based-truth, or the sense-perception-based-reality, or, if it is preferred, the sense-perception-based-evidence, the sense-perception-based-proof, or the sense-perception-based-authentication or, better still, the sense-perception-based-knowledge, the sense-perception-based-insight, or the sense-perception-based-know-how, regarding or vis-a-vis, the creator, maker, or, progenitor of the **current cosmos,** which is offered or volunteered by the consciousnessbal or the awarenessbal science of **Adwait-Vedanta** to the human consciousnesses of the **current cosmos,** is the following.

The conscious, aware, or sentient, creator, maker, or progenitor of the current cosmos, namely, the conscious, aware, or sentient plus non-physical, non-material, unphysical or immaterial, ubiquitous and infinite, 3-D or three-dimensional field of consciousness or ocean of consciousness aka the current cosmic space of mankind's fully-awake or wide-awake-state, is a soundless, silent, mum, or mute, being, truth, reality, or existence, due to the lack on its part of a physical, objective, or concrete, speech-equipment, namely, the larynx and the tongue plus the lips and the lungs. It lacks the physical, objective, or concrete, speech-equipment, namely, the larynx and the tongue plus the lips and the lungs, because, unlike the human consciousnesses of the current cosmos, it is an incredible, or awe-inspiring, the only one of its kind or unique, unembodied, disembodied, unbodied, disbodied, or bodiless consciousness. There is no one like him, her, it, or whatever in the current cosmos. That is why it is titled or tagged as 'the only one of its kind' or 'unique'.

The current **3-D** or **three-dimensional,** physical, objective, or concrete cosmos, which is floating, wafting, or levitating plus whirling, twirling, or spiralling non-stop inside the **conscious, aware,** or **sentient** plus **non-physical, non-material, unphysical,** or **immaterial,** ubiquitous and infinite, **3-D** or **three-dimensional field of consciousness** or **ocean of**

consciousness aka **cosmic space,** and has been doing so from the beginning of the **current time,** and will continue to do so till the end of the **current time,** is made or composed of, a **condensed, compressed, compacted, congealed,** or **curdled** form or version of a section, segment, part, or portion of the **conscious, aware,** or **sentient** plus **non-physical, non-material, unphysical,** or **immaterial,** ubiquitous and infinite, **3-D** or **three-dimensional, field of consciousness,** or, **ocean of consciousness,** aka the current **cosmic space** of mankind's **fully-awake** or **wide-awake-state.**

On account of the fact that, the current **3-D** or **three-dimensional,** physical, objective, or concrete cosmos is made or composed of, a **condensed, compressed, compacted, congealed,** or **curdled** form or version of a section, segment, part, or portion of the **conscious, aware,** or **sentient** plus **non-physical, non-material, unphysical,** or **immaterial,** ubiquitous and infinite, **3-D** or **three-dimensional, field of consciousness,** or, **ocean of consciousness,** aka **cosmic space,** the former, namely, the current **3-D** or **three-dimensional,** physical, objective, or concrete cosmos, is floating, wafting, or levitating plus whirling, twirling, or spiralling non-stop inside the **conscious, aware,** or **sentient** plus **non-physical, non-material, unphysical,** or **immaterial,** ubiquitous and infinite, **3-D** or **three-dimensional field of consciousness** or **ocean of consciousness,** aka the current **cosmic space** and has been doing so from the beginning of the **current time** and will continue to do so till the end of the **current time.**

Incidentally, the **current cosmos** consists not only of the countless, **3-D** or **three-dimensional,** physical, objective, or concrete moons, planets, stars, galaxies and the like but also consists of the innumerable, **3-D** or **three-dimensional,** physical, objective, or concrete bodies of the countless, manifested, incarnate, or embodied **consciousnesses** of the **current cosmos,** for example, the **3-D** or **three-dimensional,** physical, objective, or concrete, bodies of the manifested, incarnate, or embodied, human consciousnesses, of the **current cosmos.**

As said before, the conscious, aware, or sentient plus non-physical, non-material, unphysical, or immaterial, ubiquitous and infinite, 3-D or three-dimensional, field of consciousness, or, ocean of consciousness, aka cosmic space, is the creator, maker, or progenitor, of the current, 3-D or three-dimensional, physical, objective, or concrete, cosmos, consisting of the countless, 3-D or three-dimensional, physical, objective, or concrete moons, planets, stars, galaxies and the like on o hand, and the countless 3-D or three-dimensional, physical, objective, or concrete bodies of the countless, manifested, incarnate, or embodied consciousnesses of the current cosmos, on the other, for example, the 3-D or three-dimensional, physical, objective, or concrete bodies of the manifested, incarnate, or embodied, human consciousnesses of the current cosmos.

The conscious, aware, or sentient plus non-physical, non-material, unphysical, or immaterial, ubiquitous and infinite, 3-D or three-dimensional field of consciousness, or, ocean of consciousness, aka the current cosmic space of mankind's fully-awake or wide-awake-stat, is the expanded, distended, dilated, or inflated, form or version, of the dimensionless form or version, or, the original form or version, of the incredible or awe-inspiring, and the only one of its kind, or, unique, timeless, bodiless, and dimensionless, consciousness of god, creator, maker, or progenitor, of the current cosmos. The **current cosmos** is the **cosmos** which is seen, observed, and experienced by the human consciousnesses in their **fully-awake** or **wide-awake-state,** every moment of their lives.

The conscious, aware, or sentient plus non-physical, non-material, unphysical, or immaterial, ubiquitous and infinite, 3-D or three-dimensional field of consciousness, or, ocean of consciousness, aka the current cosmic space of mankind's fully-awake or wide-awake-state, has created or formed the current, 3-D or three-dimensional, physical, objective, or concrete cosmos, consisting of countless, 3-D or three-dimensional, physical, objective, or concrete moons, planets, stars, galaxies and the like on one hand, and countless 3-D or three-dimensional, physical, objective, or concrete bodies of the countless, manifested, incarnate, or embodied consciousnesses of the current cosmos, on the other, for example, the 3-D or three-dimensional, physical, objective, or concrete bodies of the manifested, incarnate, or embodied, human consciousnesses of the current cosmos, through the common-'o'-garden, or very common and ordinary activity called daydreaming, or through the common-'o'-garden, or very common and ordinary activity called oneirism, in order to amuse, entertain, or regale itself, or, if it is preferred, in order to buck up, cheer up, or buoy up itself, nothing more nothing less, because this conscious, aware, or sentient plus non-physical, non-material, unphysical, or immaterial, ubiquitous and infinite, 3-D or three-dimensional field of consciousness, or, ocean of consciousness, aka the current cosmic space of mankind's fully-awake or wide-awake-state, is absolutely lone, alone, solitary, or single, in its dimensionless form or version or, in its original form or version, and, therefore, occasionally or once in a while, it suffers from pangs, twinges, or stabs of loneliness or lonesomeness, or, if it is preferred, if suffers from pangs, twinges, or stabs of aloneness or forlornness.

When this **dimensionless** form or version i.e. when this **original** form or version of absolutely lone, alone, solitary, or single, god, creator, maker, or progenitor of the **current cosmos** suffers from pangs, twinges, or stabs of loneliness or lonesomeness, or, if it is preferred, suffers from pangs, twinges, or stabs of aloneness or forlornness, it **expands, distends, dilates,** or **inflates** its **dimensionless** form or version, or its **original** form or version, in order to form or create, enough **3-D** or **three-dimensional, consciousnessbal-space** or **awarenessbal-space,** or, if it is preferred, enough **3-D** or **three-dimensional, consciousnessbal-territory** or **awarenessbal-territory** inside itself, for the purpose of lodging, housing, or accommodating, **spatially** or **territorially,** its subsequently created, **daydream-stuff, oneiric-stuff, reverie-stuff, imagery-stuff, dreamry-stuff,** or **fantasy-stuff-composed, 3-D** or **three-dimensional,** physical, objective, or concrete **cosmos.**

As said before, this daydream-stuff, oneiric-stuff, reverie-stuff, imagery-stuff, dreamry-stuff, or fantasy-stuff-composed, 3-D or three-dimensional, physical, objective, or concrete cosmos is created by this god, creator, maker, or progenitor of the current cosmos, through the common-'o'-garden, or very common and ordinary activity called daydreaming, or

through the common-'o'-garden, or very common and ordinary activity called oneirism on its part, nothing more nothing less.

Hence, the current **3-D** or **three-dimensional,** physical, objective, or concrete cosmos, consisting of countless, **3-D** or **three-dimensional,** physical, objective, or concrete moons, planets, stars, galaxies and the like, on one hand, and countless **3-D** or **three-dimensional,** physical, objective, or concrete bodies of the countless, manifested, incarnate, or embodied **consciousnesses** of the **current cosmos,** on the other, for example, the **3-D** or **three-dimensional,** physical, objective, or concrete bodies of the manifested, incarnate, or embodied, human consciousnesses of the **current cosmos,** are made or composed of, nothing more than the **daydream-stuff, oneiric-stuff, reverie-stuff, imagery-stuff, dreamry-stuff,** or **fantasy-stuff** of no less a being than the **conscious, aware,** or **sentient** plus **non-physical, non-material, unphysical,** or **immaterial,** ubiquitous and infinite, **3-D** or **three-dimensional field of consciousness** or **ocean of consciousness** aka the current **cosmic space** which, in turn, is nothing but the **expanded, distended, dilated,** or **inflated** form or version of the **dimensionless** form or version or the **original** form or version of the incredible or awe-inspiring, and **the only one of its kind** or unique, **timeless, bodiless,** and **dimensionless consciousness** of god, creator, maker, or progenitor of the current cosmos.

Incidentally, the current cosmos is the cosmos, which is seen, observed, and experienced by the human consciousnesses in their **fully-awake** or **wide-awake-state** every moment of their lives.

COSMIC SPACE IS THE EXPANDED, VISIBLE, AND THREE-DIMENSIONAL EXPRESSION OF THE DIMENSIONLESS CONSCIOUSNESS OF GOD

The current **timeless, 3-D** or **three-dimensional,** non-physical, non-material, unphysical, or immaterial **cosmic space** is beheld, discerned, seen or viewed by the **timeless, dimensionless,** non-physical, non-material, unphysical, or immaterial **embodied** human consciousnesses of the **current cosmos** in their fully-awake or wide-awake state through the medium of the **time-bound** and **3-D** or **three-dimensional,** insentient or incapable of feeling or understanding **physical** eyes of their **time-bound** and **3-D** or **three-dimensional,** insentient or incapable of feeling or understanding **concrete, objective** or **material** body.

This **timeless, 3-D** or **three-dimensional,** non-physical, non-material, unphysical, or immaterial **cosmic space** is not an insentient or incapable of understanding or feeling thing or entity which just happens to be in existence in the **current cosmos** accidentally, by chance, by serendipity, or by fluke, or, better still, which has managed to come into being in the **current cosmos** or which has managed to arrive, emerge or surface in the **current cosmos** accidentally, by chance, or by happy coincidence from nowhere for the purpose of spatially accommodating all the current **time-bound** and **3-D** or **three-dimensional,** concrete-matter-composed, objective-matter-composed, or physical-matter-composed, insentient or incapable of feeling or understanding countless things or items of the **current cosmos,** for example, all the current **time-bound** and **3-D** or **three-dimensional,** concrete-matter-composed, objective-matter-composed, or physical-matter-composed, insentient or one hundred percent incapable of feeling or understanding moons, planets, stars, galaxies and the like on one hand, and all the current **time-bound** and **3-D** or **three-dimensional,** concrete-matter-composed, objective-matter-composed, or physical-matter-composed, insentient or one hundred percent incapable of feeling or understanding bodies or anatomies of all the **timeless, dimensionless,** non-physical, non-material, unphysical, or immaterial **embodied** consciousnesses of the **current cosmos** on the other, for example, all the **time-bound** and **3-D** or **three-dimensional,** concrete-matter-composed, objective-matter-composed, or physical-matter-composed, insentient or one hundred percent incapable of feeling or understanding bodies or anatomies of all the **timeless, dimensionless,** non-physical, non-material, unphysical, or immaterial **embodied** human consciousnesses, awarenesses, sentiences, minds or 'I's of the **current cosmos.**

Instead, the current **timeless, 3-D** or **three-dimensional,** non-physical, non-material, unphysical, or immaterial **cosmic space** is something quite different.

To put it very honestly and candidly, the current **timeless, 3-D** or **three-dimensional,** non-physical, non-material, unphysical, or immaterial **cosmic space;** which is beholdable, discernible, seeable, viewable or visible to all the **timeless** and **dimensionless,** non-physical, non-material, unphysical, or immaterial **embodied** human consciousnesses of the **current cosmos** in their fully-awake or wide-awake state, through the medium of the **time-bound** and **3-D** or **three-dimensional, physical** eyes of their **time-bound** and **3-D** or **three-dimensional, objective, concrete** or **material** body; is something which is supremely vital or essential for the birth, creation or genesis of the current **time-bound, 3-D** or **three-dimensional,** insentient or incapable of feeling or understanding physical, material, objective, or concrete cosmos.

The current **time-bound, 3-D** or **three-dimensional,** insentient or one hundred incapable of feeling or understanding **physical, material, objective** or **concrete** cosmos consists of countless insentient or one hundred percent incapable of feeling and understanding **time-bound** and **3-D** or **three-dimensional,** physical, material, objective or concrete things, beings or entities of the current cosmos, for example, all the moons, planets, stars, galaxies and the like of the current cosmos on one hand, and all the **time-bound** and **3-D** or **three-dimensional,** physical, material, objective or concrete bodies or anatomies of all the **timeless, dimensionless,** non-physical, non-material, unphysical or immaterial **embodied** consciousnesses of the current cosmos on the other, for example, all the **time-bound** and **3-D** or

three-dimensional, physical, material, objective or concrete bodies or anatomies of all the **timeless, dimensionless,** non-physical, non-material, unphysical or immaterial **embodied** human consciousnesses of the current cosmos.

All the above mentioned, **time-bound, 3-D** or **three-dimensional,** insentient or one hundred percent incapable of feeling or understanding **physical, material, objective** or **concrete** things, beings or items of the current **time-bound** and **3-D** or **three-dimensional,** insentient or one hundred percent incapable of feeling or understanding **physical, material, objective** or **concrete** cosmos, are floating, wafting or levitating plus whirling, twirling or spiraling non-stop inside the current **timeless** and **3-D** or **three-dimensional,** non-physical, non-material, unphysical, or immaterial **cosmic space** and has been doing so from the beginning of the current time and will continue to do so till the end of the current time.

What has been said above with regards to the current **time-bound, 3-D** or **three-dimensional,** insentient or one hundred percent incapable of feeling or understanding **physical, material, objective** or **concrete** cosmos, can be put in another way.

The current **time-bound, 3-D** or **three-dimensional,** insentient or one hundred percent incapable of feeling or understanding **physical, material, objective** or **concrete** cosmos is the domain, of which the current **timeless, dimensionless,** non-physical, non-material, unphysical, or immaterial **embodied** human consciousnesses, are a constituent part.

All the **timeless, dimensionless,** non-physical, non-material, unphysical, or immaterial **embodied** human consciousnesses of the current cosmos; which presently are a **constituent part** of the current **time-bound, 3-D** or **three-dimensional,** insentient or one hundred percent incapable of feeling or understanding **physical, material, objective or concrete** cosmos; observe, perceive and experience in their **fully-awake** or **wide-awake-state,** the current **time-bound, 3-D** or **three-dimensional,** insentient or one hundred percent incapable of feeling or understanding **physical, material, objective** or **concrete** cosmos, not only through the medium of the **time-bound** and **3-D** or **three-dimensional,** insentient or one hundred percent incapable of feeling or understanding **physical** eyes of their **time-bound** and **3-D** or **three-dimensional,** insentient or one hundred percent incapable of feeling or understanding **physical, material, objective** or **concrete** body, but also through the media of all the other **time-bound** and **3-D** or **three-dimensional,** insentient or one hundred percent incapable of feeling or understanding **physical** senses of their **time-bound** and **3-D** or **three-dimensional,** insentient or one hundred percent incapable of feeling or understanding **physical, material, objective** or **concrete** body.

Let one now explain, what the current **timeless, 3-D** or **three-dimensional,** non-physical, non-material, unphysical, or immaterial, **cosmic space** truly is.

As said before, the current **timeless, 3-D** or **three-dimensional,** non-physical, non-material, unphysical, or immaterial, **cosmic space** is beholdable, detectable, discernible, noticeable, seeable, viewable or visible to all the **timeless, dimensionless,** non-physical, non-material, unphysical, or immaterial **embodied** human consciousnesses of the **current cosmos** in their fully-awake or wide-awake state, through the medium of the **time-bound, 3-D** or **three-dimensional,** insentient or one hundred percent incapable of feeling or understanding **physical** eyes of their **time-bound, 3-D** or **three-dimensional,** insentient or one hundred percent incapable of feeling or understanding **objective, concrete** or **material** body.

The task of explaining, what the current **timeless, 3-D** or **three-dimensional,** non-physical, non-material, unphysical, or immaterial **cosmic space** truly is, will be best achieved by first bringing to the notice of the **timeless, dimensionless,** non-physical, non-material, unphysical, or immaterial, **embodied** human consciousness of the **current cosmos,** the fact of existence of one of their very common-'o'-garden or ordinary adventure or exploit, called the adventure or exploit of daydreaming or oneirism, or, if it is preferred, the adventure or exploit of dreamism, reverism, reverie, dreamry or imagery, or, better still, the adventure or exploit of building or forming the insentient or one hundred percent incapable of feeling or understanding, **time-bound** and **3-D-looking** or **three-dimensional-looking** plus **physical-looking, objective-looking,** or **concrete-looking, daydream-stuff-composed, oneiric-stuff-composed, dreamry-stuff-composed, reverie-stuff-composed,** or **imagery-stuff-composed** cosmoses, castles or whatever inside themselves or, if it is preferred inside their **timeless, dimensionless,** non-physical, non-material, unphysical or immaterial **embodied** consciousness, awarenesses, sentiences, minds or Is.

These insentient or one hundred percent incapable of feeling or understanding, time-bound and 3-D-looking or three-dimensional-looking plus physical-looking, objective-looking, or concrete-looking, daydream-stuff-composed, oneiric-stuff-composed, dreamry-stuff-composed, reverie-stuff-composed, or imagery-stuff-composed cosmoses, castles or whatever are built or formed by all the timeless, dimensionless, non-physical, non-material, unphysical, or immaterial, embodied human consciousness of the current cosmos inside their timeless, dimensionless, non-physical, non-material, unphysical, or immaterial embodied consciousness during their activity of daydreaming, oneirism, dreamism, reverism, reverie, dreamry or imagery in their fully-awake or wide-awake state.

This adventure or exploit of daydreaming, oneirism, dreamism, reverism, reverie, dreamry or imagery is accomplished by the **timeless, dimensionless,** non-physical, non-material, unphysical, or immaterial **embodied** human consciousnesses in their **fully-awake** or **wide-awake-state** when they daydream, oneirate, reverie, fantasy, conceive, envisage, ideate, imagine, or fantasize inside or within their own consciousnesses or minds, that is to say, when they muse, or give free rein to their imagination to build or form insentient or one hundred percent incapable of feeling or understanding, cosmoses, castles or whatever inside their own consciousnesses or minds in their **fully-awake** or **wide-awake** state.

What has been said above can be put in another way.

The adventure or exploit of daydreaming, oneirism, dreamism, reverism, reverie, dreamry or imagery is accomplished by the **timeless, dimensionless,** non-physical, non-material, unphysical, or immaterial **embodied** human consciousnesses in their **fully-awake** or **wide-awake-state** when they form or create, an insentient or one hundred percent incapable of feeling or understanding **time-bound** and **3-D** or **three-dimensional,** daydream-stuff-composed, oneiric-stuff-composed, dreamry-stuff-composed, reverie-stuff-composed or imagery-stuff-composed cosmoses or castles or whatever inside themselves or, if it is preferred, inside their **timeless, dimensionless,** non-physical, non-material, unphysical, or immaterial **embodied** consciousnesses in their **fully-awake** or **wide-awake** state.

In other words, the adventure or exploit of daydreaming, oneirism, dreamism, reverism, reverie, dreamry, or imagery is accomplished by the **timeless, dimensionless,** non-physical, non-material, unphysical, or immaterial, **embodied** human consciousnesses in their **fully-awake** or **wide-awake-state,** when they build or form inside themselves or, if it is preferred, when they build or form inside their **timeless, dimensionless,** non-physical, non-material, unphysical, or immaterial **embodied** consciousnesses or minds an insentient or one hundred percent incapable of feeling or understanding **time-bound, 3-D-looking** or **three-dimensional-looking** plus **physical-looking, objective-looking,** or **concrete-looking,** consciousnessbal-stuff-composed, awarenessbal-stuff-composed, sentiences-stuff-composed, mind-stuff-composed or I-stuff-composed cosmoses or castles or whatever in their **fully-awake** or **wide-awake state.**

The above mentioned, insentient or one hundred percent incapable of feeling or understanding **time-bound, 3-D-looking** or **three-dimensional-looking,** daydream-stuff-composed, oneiric-stuff-composed, dreamry-stuff-composed, reverie-stuff-composed, or imagery-stuff-composed cosmoses or castles or whatever, or, if it is preferred, the above mentioned, insentient or one hundred percent incapable of feeling or understanding **time-bound, 3-D-looking** or **three-dimensional-looking** plus **physical-**

looking, objective-looking, or **concrete-looking, consciousnessbal-stuff-composed, awarenessbal-stuff-composed, sentiencel-stuff-composed, mind-stuff-composed** or **I-stuff-composed** cosmoses or castles or whatever, built or formed by the **timeless, dimensionless,** non-physical, non-material, unphysical, or immaterial, **embodied** human consciousnesses inside themselves, in their **fully-awake** or **wide-awake-state,** through their very common-'o'-garden or ordinary activity of daydreaming or oneirism, or, if it is preferred, through their very common-'o'-garden or ordinary activity of dreamism, reverism, reverie, dreamry, or imagery, or, better still, through their very common-'o'-garden or ordinary activity of building or forming insentient or one hundred percent incapable of feeling or understanding **time-bound, 3-D-looking** or **three-dimensional-looking** plus **physical-looking, objective-looking,** or **concrete-looking, daydream-stuff-composed, oneiric-stuff-composed, dreamry-stuff-composed, reverie-stuff-composed,** or **imagery-stuff-composed,** cosmoses or castles or whatever inside themselves or, better still, inside their **timeless, dimensionless,** non-physical, non-material, unphysical, or immaterial **embodied** consciousnesses or minds, can be as varied, diversified, or wide-ranging as these **timeless, dimensionless,** non-physical, non-material, unphysical, or immaterial, **embodied** human consciousness or minds of the **current cosmos** in their fully-awake-awake or wide-awake-state wish, want, crave or desire.

During their activity of daydreaming, or oneirism, or, if it is preferred, during their activity of dreamism, reverism, reverie, dreamry, or imagery, or, better still, during their activity of dream-making, oneiric-making, dreamry-making, reverie-making or imagery-making inside themselves, or, better still, inside their **timeless, dimensionless,** non-physical, non-material, unphysical, or immaterial, **embodied** consciousnesses or minds, the latter, namely, the **timeless, dimensionless,** non-physical, non-material, unphysical, or immaterial **embodied** human consciousnesses or minds, inevitably, perforce, or, of necessity, have to expand, distend, inflate, or dilate themselves or, if it is preferred, have to expand, distend, dilate, or inflate their **timeless, dimensionless,** non-physical, non-material, unphysical, or immaterial, **embodied** consciousnesses or minds, in order to create a **timeless, 3-D** or **three-dimensional,** non-physical, non-material, unphysical, or immaterial **cosmic-space** inside themselves, a **timeless, 3-D** or **three-dimensional,** non-physical, non-material, unphysical, or immaterial **cosmic-space,** which is nothing but a **timeless, 3-D** or **three-dimensional,** non-physical, non-material, unphysical, or immaterial **consciousnessbal-space, awarenessbal-space, sentiencel-space, mind-space,** or **I-space** situated or located inside their expanded, distended, dilated, or inflated, **timeless, 3-D** or **three-dimensional,** non-physical, non-material, unphysical, or immaterial **embodied, consciousnesses, awarenesses, sentiences, minds,** or '**I's** and nowhere else.

To repeat.

During their activity of daydreaming, or oneirism, or, if it is preferred, during their activity of dreamism, reverism, reverie, dreamry, or imagery, or, better still, during their activity of dream-making, oneiric-making, dreamry-making, reverie-making or imagery-making inside themselves, or, better still, inside their **timeless, dimensionless,** non-physical, non-material, unphysical, or immaterial **embodied** consciousnesses, the latter, namely, the **timeless,**

dimensionless, non-physical, non-material, unphysical, or immaterial **embodied** human consciousnesses or minds, inevitably, perforce, or, of necessity have to expand, distend, inflate, or dilate themselves or, if it is preferred, have to expand, distend, dilate, or inflate their **timeless, dimensionless,** non-physical, non-material, unphysical, or immaterial, **embodied** consciousnesses or minds, in order to create a **timeless, 3-D** or **three-dimensional,** non-physical, non-material, unphysical, or immaterial, **cosmic-space** inside themselves, a **timeless, 3-D** or **three-dimensional,** non-physical, non-material, unphysical, or immaterial, **cosmic-space,** which is nothing but a **timeless, 3-D** or **three-dimensional,** non-physical, non-material, unphysical or immaterial, **consciousnessbal-space, awarenessbal-space, sentiencel-space, mind-space,** or **I-space,** situated or located inside their expanded, distended, dilated, or inflated **consciousnesses, awarenesses, sentiences, minds,** or **'I's** and nowhere else.

That is to say, **timeless, dimensionless,** non-physical, non-material, unphysical, or immaterial, **embodied** human consciousnesses or minds, inevitably, perforce, or, of necessity have to expand, distend, inflate, or dilate themselves or, if it is preferred, have to expand, distend, dilate, or inflate their **timeless, dimensionless,** non-physical, non-material, unphysical, or immaterial, **embodied** consciousnesses or minds, during their activity of daydreaming, or oneirism, or, if it is preferred, during their activity of dreamism, reverism, reverie, dreamry, or imagery, or, better still, during their activity of daydream-making, oneiric-making, dreamry-making, reverie-making, or imagery-making, inside themselves, or, if it is preferred, inside their **timeless, dimensionless,** non-physical, non-material, unphysical, or immaterial **embodied** consciousnesses or minds, in order to form or create a **timeless, 3-D** or **three-dimensional,** non-physical, non-material, unphysical, or immaterial **cosmic-space** inside themselves, a **timeless, 3-D** or **three-dimensional,** non-physical, non-material, unphysical, or immaterial **cosmic-space,** which is nothing but a **timeless, 3-D** or **three-dimensional,** non-physical, non-material, unphysical, or immaterial **consciousnessbal-space, awarenessbal-space, sentiencel-space, mind-space,** or **I-space,** situated or located inside their expanded, distended, dilated, or inflated, **timeless, 3-D** or **three-dimensional,** non-physical, non-material, unphysical, or immaterial **embodied** consciousnesses, awarenesses, sentiences, minds or 'I's and nowhere else.

What has been said above can be put in another way.

The **timeless, dimensionless,** non-physical, non-material, unphysical, or immaterial, **embodied** human consciousnesses or minds, inevitably, perforce, or, of necessity have to expand, distend, dilate or inflate themselves or, if it is preferred, have to expand, distend, dilate, or inflate their **timeless, dimensionless,** non-physical, non-material, unphysical, or immaterial, **embodied** consciousnesses or minds in order to form or create a **timeless, 3-D** or **three-dimensional,** non-physical, non-material, unphysical, or immaterial, **cosmic-space-mimicking, cosmic-space-resembling,** or **cosmic-space-imitating,** consciousnessbal-space or mind-space inside themselves with a view to spatially accommodate all their insentient or incapable of feeling or understanding **time-bound, 3-D-looking** or **three-dimensional-looking** plus **physical-looking, objective-looking,** or **concrete-looking,** daydream-stuff-composed, oneiric-stuff-composed, dreamry-stuff-composed, reverie-stuff-composed, or imagery-stuff-composed cosmoses, castles or whatever inside themselves or, if it is preferred, inside their **timeless, dimensionless,** non-physical, non-material, unphysical, or immaterial, **embodied** consciousnesses, awarenesses, sentience, minds or Is.

With regards to comprehending the true nature of the current **timeless, 3-D** or **three-dimensional,** non-physical, non-material, unphysical or immaterial **cosmic space** inside which the current **time-bound, 3-D** or **three-dimensional,** insentient or incapable of feeling or understanding physical, objective, or concrete cosmos of mankind' **fully-awake** or **wide-awake-state** is floating, wafting, or levitating plus whirling, twirling, or spiraling non-stop and has been doing so from the beginning of current time and will continue to do so till the end of the current time, the **timeless, dimensionless,** non-physical, non-material, unphysical, or immaterial **embodied** human consciousnesses must take a clue or hint from all that has been said above.

Let one elaborate.

With regards to comprehending the true nature of the current **timeless, 3-D** or **three-dimensional,** non-physical, non-material, unphysical or immaterial **cosmic space** inside which the current **time-bound, 3-D** or **three-dimensional,** insentient or incapable of feeling or understanding physical, objective, or concrete cosmos of mankind's **fully-awake** or **wide-awake-state,** is floating, wafting, or levitating plus whirling, twirling, or spiraling non-stop and has been doing so from the beginning of current time and will continue to do so till the end of the current time, the **timeless, dimensionless,** non-physical, non-material, unphysical, or immaterial **embodied** human consciousnesses of the **current cosmos** must take a clue or hint from all that has been said above vis-a-vis, about or concerning the **timeless, dimensionless,** non-physical, non-material, unphysical, or immaterial, **embodied** human consciousnesses of the current cosmos but purely or specifically with respect to their innate ability to form or create inside themselves or, better still, but purely or specifically with respect to their innate ability to form or create inside their **timeless, dimensionless,** non-physical, non-material, unphysical, or immaterial **embodied** consciousnesses, a **time-bound, 3-D** or **three-dimensional,** non-physical, non-material, unphysical, or immaterial, **cosmic-space-mimicking, cosmic-space-resembling,** or **cosmic-space-imitating,** consciousnessbal-space or mind-space, with a view to spatially accommodate all their **time-bound, 3-D-looking** or **three-dimensional-looking** plus **physical-looking, objective-looking,** or **concrete-looking,** daydream-stuff-composed, oneiric-stuff-composed, dreamry-stuff-composed, reverie-stuff-composed, or imagery-stuff-composed, insentient or incapable of feeling or understanding cosmoses, castles or whatever, which their **timeless, dimensionless,** non-physical, non-material, unphysical, or immaterial, **embodied** consciousnesses have formed or created inside themselves through the very common-'o'-garden or ordinary activity of daydreaming or oneirism on their part, or, if it is preferred, through the very common-'o'-garden or ordinary activity of dreamism, reverism, reverie, dreamry or imagery on their part, or, still better, through the very common-'o'-garden or ordinary activity of dream-making, oneiric-making, dreamry-making, reverie-making, or imagery-making, on their part.

As said before, the current **cosmic space** is a **timeless, 3-D** or **three-dimensional,** non-physical, non-material, unphysical or immaterial **being** which the **timeless, dimensionless,** non-physical, non-material, unphysical or immaterial, **embodied** human consciousnesses of the **current cosmos** behold, detect, discern, notice, see or view in their **fully-awake** or **wide-awake-state** through the medium of the **time-bound, 3-D** or **three-dimensional,** insentient or incapable of feeling or understanding, **physical** eyes of their **time-bound, 3-D** or **three-dimensional,** insentient or incapable of feeling or understanding **objective, concrete** or **material** body or anatomy.

It is inside this **timeless, 3-D** or **three-dimensional,** non-physical, non-material, unphysical or immaterial **cosmic space** that all the current **time-bound, 3-D** or **three-dimensional,** insentient or one hundred percent incapable of feeling or understanding physical, material, objective or concrete things or items of the current cosmos, for example, all the current **time-bound, 3-D** or **three-dimensional,** insentient or one hundred percent incapable of feeling or understanding physical, material, objective or concrete moons, planets, stars, galaxies and the like on one hand, and all the current **time-bound, 3-D** or **three-dimensional,** insentient or one hundred percent incapable of feeling or understanding physical, material, objective or concrete bodies or anatomies of all the current **timeless, dimensionless,** non-physical, non-material, unphysical or immaterial **embodied** consciousnesses of the current cosmos on the other, for example, all the current **time-bound, 3-D** or **three-dimensional,** physical, material, objective or concrete bodies or anatomies of all the current **timeless, dimensionless,** non-physical, non-material, unphysical or immaterial **embodied** human consciousnesses of the current cosmos, are floating, wafting or levitating plus whirling, twirling or spiraling non-stop and has been doing so from the beginning of the current time and will continue to do so till the end of the current time.

If the **timeless, dimensionless,** non-physical, non-material, unphysical or immaterial **embodied** human consciousnesses of the **current cosmos** crave, yearn or pine for comprehending the true nature of the current **timeless, 3-D** or **three-dimensional,** non-physical, non-material, unphysical or immaterial **cosmic space,** then they must seize or snatch the following truth about this **cosmic space** from all that has been said above.

Just to recapitulate or summarize that the current **cosmic space** is the incredible **timeless, 3-D** or **three-dimensional,** non-physical, non-material, unphysical or immaterial **being** which the current **timeless, dimensionless,** non-physical, non-material, unphysical or immaterial **embodied** human consciousnesses behold, discern, notice, see or view in their **fully-awake** or **wide-awake-state** through the medium of the **3-D** or **three-dimensional,** insentient or incapable of feeling or understanding **physical** eyes of their **3-D** or **three-dimensional,** insentient or incapable of feeling or understanding, **objective, concrete** or **material** body or anatomy and inside which all the current **time-bound, 3-D** or **three-dimensional,** insentient or one hundred percent incapable of feeling or understanding physical, material, objective or concrete things or items of the **current cosmos,** for example, all the current **time-bound, 3-D** or **three-dimensional,** insentient or one hundred percent incapable of feeling or understanding physical, material, objective or concrete moons, planets, stars, galaxies and the like on one hand, and all the current **time-bound, 3-D** or **three-dimensional,** insentient or one hundred percent incapable of feeling or understanding physical, material, objective or concrete bodies or anatomies of all the current **timeless, dimensionless,** non-physical, non-material, unphysical or immaterial **embodied** consciousnesses of the current cosmos on the other, for example, all the current **time-bound, 3-D** or **three-dimensional,** physical, material, objective or concrete bodies or anatomies of all the current **timeless, dimensionless,** non-physical, non-material, unphysical or immaterial **embodied** human consciousnesses of the current cosmos, are floating, wafting or levitating plus whirling, twirling or spiraling non-stop and has been doing so from the beginning of the current time and will continue to do so till the end of the current time.

In terms of absolute truth, the current timeless, 3-D or three-**dimensional,** non-physical, non-material, unphysical, or immaterial **cosmic space,** beheld, discerned, noticed, seen or visualized by human consciousnesses in their **fully-awake** or **wide-awake-state,** is the **expanded, distended, dilated,** or **inflated** form or version of the incredible or awe-inspiring, **the only one of its kind** or unique, **timeless, bodiless,** and **dimensionless** consciousness of god, creator, maker, or progenitor of the current **time-bound, 3-D** or **three-dimensional,** insentient or incapable of feeling or understanding, **physical, material, objective** or **concrete** cosmos, nothing more nothing less.

To repeat.

If the **timeless, dimensionless,** non-physical, non-material, unphysical or immaterial **embodied** human consciousnesses of the **current cosmos** crave, yearn or pine for comprehending the true nature of the current **timeless, 3-D** or **three-dimensional,** non-physical, non-material, unphysical or immaterial **cosmic space,** then they must seize or snatch the following truth about this **cosmic space** from all that has been said above.

Just to recapitulate or summarize that the current **cosmic space** is the incredible **timeless, 3-D** or **three-dimensional,** non-physical, non-material, unphysical or immaterial **being** which the current **timeless, dimensionless,** non-physical, non-material, unphysical or immaterial **embodied** human consciousnesses behold, discern, notice, see or view in their **fully-awake** or **wide-awake-state** through the medium of the **3-D** or **three-dimensional,** insentient or incapable of feeling or understanding **physical** eyes of their **3-D** or **three-dimensional,** insentient or incapable of feeling or understanding **objective, concrete** or **material** body or anatomy and inside which all the current **time-bound, 3-D** or **three-dimensional,** insentient or one hundred percent incapable of feeling or understanding, physical, material, objective or concrete things or items of the **current cosmos,** for example, all the current **time-bound, 3-D** or **three-dimensional,** insentient or one hundred percent incapable of feeling or understanding physical, material, objective or concrete moons, planets, stars, galaxies and the like on one hand, and all the current **time-bound, 3-D** or **three-dimensional,** insentient or one hundred percent incapable of feeling or understanding physical, material, objective or concrete bodies or anatomies of all the current **timeless, dimensionless,** non-physical, non-material, unphysical or immaterial **embodied** consciousnesses of the current cosmos on the other, for example, all the current **time-bound, 3-D** or **three-dimensional,** physical, material, objective or concrete bodies or anatomies of all the current **timeless, dimensionless,** non-physical, non-material, unphysical or immaterial **embodied** human consciousnesses of the current cosmos, are floating, wafting or levitating plus whirling, twirling or spiraling non-stop and has been doing so from the beginning of the current time and will continue to do so till the end of the current time.

In terms of absolute truth, the current timeless, 3-D or three-**dimensional,** non-physical, non-material, unphysical, or immaterial **cosmic space,** is the **expanded, distended, dilated,** or **inflated** form or version of the incredible or awe-inspiring, **the only one of its kind** or unique, **timeless, bodiless,** and **dimensionless** consciousness of god, creator, maker, or progenitor of the current, **time-bound, 3-D** or **three-dimensional,** insentient or incapable of feeling or understanding **physical, material, objective** or **concrete** cosmos, nothing more nothing less.

Just to remind oneself that the current **cosmic space** is the incredible, **timeless, 3-D** or **three-dimensional,** non-physical, non-material, unphysical, or immaterial **being** which is beheld, discerned, noticed, seen or visualized by all the **timeless, dimensionless,** non-physical, non-material, unphysical, or immaterial, **embodied** human consciousnesses of the **current cosmos** in their fully-awake or wide-awake state through the medium of the **time-bound, 3-D** or **three-dimensional,** insentient or incapable of feeling or understanding **physical** eyes of their **time-bound, 3-D** or **three-dimensional,** insentient or incapable of feeling or understanding **objective, concrete** or **material** body or anatomy.

The current **cosmic space** is also that incredible or extraordinary, **timeless, 3-D** or **three-dimensional,** non-physical, non-material, unphysical, or immaterial **being** inside which all the current insentient or one hundred percent incapable of feeling or understanding plus **time-bound, 3-D** or **three-dimensional,** physical, material, objective or concrete things or items of the current cosmos, for example, all the current insentient or one hundred percent incapable of feeling or understanding, physical, material, objective or concrete moons, planets, stars, galaxies and the like on one hand, and all the current, insentient or one hundred percent incapable of feeling or understanding, physical, material, objective or concrete bodies or anatomies of all the **timeless, bodiless,** non-physical, non-material, unphysical or immaterial **embodied** consciousnesses of the current cosmos on the other, for example, all the current insentient or one hundred percent incapable of feeling or understanding, physical, material, objective or concrete bodies or anatomies of all the **embodied** human consciousnesses of the current cosmos, are floating, wafting or levitating plus whirling, twirling or spiraling non-stop and has been doing so from the beginning of the current time and will continue to do so till the end of the current time.

To repeat.

The current **cosmic space** is the incredible, **timeless, 3-D** or **three-dimensional,** non-physical, non-material, unphysical, or immaterial **being** which all the **timeless, dimensionless,** non-physical, non-material, unphysical, or immaterial **embodied** human consciousnesses of the **current cosmos** behold, discern, notice, see or visualize through the medium of the **time-bound** and **3-D** or **three-dimensional** physical eyes of their insentient or incapable of feeling or understanding, material, objective, or concrete, **time-bound, 3-D** or **three-dimensional** body or anatomy in their **fully-awake** or **wide-awake-state,** every moment of their existence, presence, isness, or beingness in the **current cosmos.**

Furthermore, the current **cosmic space** is that incredible or extraordinary, **timeless, 3-D** or **three-dimensional,** non-physical, non-material, unphysical, or immaterial **being** inside which the current **time-bound, 3-D** or **three-dimensional,** insentient or incapable of feeling or understanding physical, objective, or concrete cosmos, consisting of countless, **time-bound, 3-D** or **three-dimensional,** insentient or incapable of feeling or understanding physical, objective, or concrete moons, planets, stars, galaxies and the like on one hand, and all the countless, **time-bound, 3-D** or **three-dimensional,** insentient or incapable of feeling or understanding physical, objective, or concrete bodies or anatomies of all the countless, **timeless, dimensionless,** non-physical, non-material, unphysical or immaterial **embodied,** consciousnesses of the **current cosmos,** on the other, for example, all the **time-bound, 3-D** or **three-dimensional,** insentient or incapable of feeling or understanding physical, objective, or concrete bodies or anatomies of all the **embodied** human consciousnesses of the **current cosmos** are floating, wafting, or levitating plus whirling, twirling, or spiraling non-stop and has been doing so from the beginning of the **current time** and will continue to do so till the end of the **current time.**

WHY GOD GAVE HUMAN CONSCIOUSNESS THE ABILITY TO DAYDREAM

The **current cosmos,** which includes its most accessible entity called the **physical matter,** is observed, perceived and experienced by all the **embodied** human consciousnesses of the **cosmos** in their **fully-awake** or **wide-awake-state ,** is a very baffling **truth** or is an outstandingly **'hard nut to crack'** in all respects, for example, to name just a few, its **source,** the **reason for its being,** and the **true nature** of its **three fundamental constituents** namely, the **cosmic space** on one hand and its myriad **embodied** consciousnesses including the **embodied** human consciousnesses on the other, not forgetting its insentient or incapable of feeling or understanding **physical matter** which composes all the moons, planets, stars and galaxies of the **current cosmos** on one hand and all the physical bodies of all the **embodied** consciousnesses of the **current cosmos** on the other, for example, all the physical bodies of all the **embodied** human consciousnesses of the **current cosmos.**

Not many **embodied** human consciousnesses of the **current cosmos** are interested in the above questions pertaining to the **current cosmos** because god, creator, maker, progenitor or whatever of the **current cosmos** has **willed** that it will be so. As a result, the great majority of **embodied** human consciousnesses of the **current cosmos** get on with their daily lives as best as it is possible for them to do and they continue to do this till the last moment of their **temporary existence** in this amazing **current cosmos.**

However, as per the **will** of god, creator, maker, progenitor or whatever of the **current cosmos,** a few amongst the great bulk of **embodied** human consciousnesses of the **current cosmos** become interested in finding the answers to the above questions pertaining to the **current cosmos.** For the assistance of these **embodied** human consciousnesses, god, creator, maker, progenitor or whatever of the **current cosmos** has provided some clues so that they may be able to work out the answers with regards to all the above questions pertaining to the **current cosmos.**

One such assistance provided by god, creator, maker, progenitor or whatever of the **current cosmos** to these **embodied** human consciousnesses of the **current cosmos** in this connection is the **universal ability** of all the **embodied** human consciousnesses to **daydream.** This **universal ability** of all the **embodied** human consciousnesses to **daydream** is the **most importance** assistance provided to these **embodied** human consciousnesses by god, creator, maker, progenitor or whatever of the **current cosmos** because it is through the instrumentality of this process called **daydreaming** that god, creator, maker, progenitor or whatever of the **current cosmos** has made, created or formed the **current cosmos** of mankind's **fully-awake** or **wide-awake-state.**

IF A STATED TRUTH IS REALLY ABSOLUTE IT WILL BE SIMPLE TO GRASP AND ACCESSIBLE TO ALL

The **current cosmos** is composed of two **basic ingredients** namely, the **concrete** or the **physical** on one hand, and the **non-concrete** or the **non-physical** on the other.

The **concrete** or the **physical aspect** of the **current cosmos** is very patent, evident, noticeable or tangible to all the **concretely** or **physically** embodied or anatomied **human** consciousnesses of the **current cosmos** because this **aspect** is available to them for the **concrete** or **physical** sense-perception through the medium of the **concrete** or the **physical** sense organs of their **concrete** or the **physical** bodies or anatomies.

In contrast, the **non-concrete** or the **non-physical** aspect of the **current cosmos,** namely the **cosmic space** on one hand and all the **concretely** or **physically** embodied or anatomied consciousnesses of the **current cosmos** on the other, for example, all the **concretely** or **physically** embodied or anatomied **human** consciousnesses of the **current cosmos**, are quite different from the **concrete** or the **physical** aspect of the **current cosmos** in many respects. To name just one of these differences, for example, is the fact of amenableness of the latter, namely the **physical** aspect, to the **concrete** or **physical** sense-perception.

Let one explain.

Whereas the **concrete** or the **physical** aspect of the **current cosmos** is amenable to **concrete** or **physical** sense-perception through the medium of the **concrete** or the **physical** sense-organs of the **concrete** or the **physical** bodies or anatomies of all the **concretely** or **physically** embodied or anatomied consciousnesses of the **current cosmos,** for example, **human** consciousnesses of the **current cosmos,** this is absolutely not possible in the case of or vis-a-vis the consciousnesses of the **current cosmos** namely the **concretely** or **physically** embodied or anotomied consciousnesses of the **current cosmos,** for example, the **concretely** or **physically** embodied or anatomied **human** consciousnesses of the **current cosmos.**

Even in the case of **cosmic space** of the **current cosmos,** the **concrete** or the **physical** sense-perception is possible only through the medium of the **concrete** or the **physical** eyes of the **concretely** or **physically** embodied or anatomied consciousnesses of the **current cosmos,** for example, the **concrete** or the **physical** eyes of the **concretely** or **physically** embodied or anatomied **human** consciousnesses of the **current cosmos.**

With the sole exception of **visual** or **ocular** sense-perception through the medium of the **concrete** or the **physical** eyes, the **concrete** or the **physical** sense-perception is not possible through the medium of any other **concrete** or the **physical** sense organ, in the case of or vis-a-vis the **cosmic space** of the **current cosmos.**

Having brought to the notice to oneself the fact of existence or presence of the two **basic ingredients** in the **current cosmos** and the differences that obtain between them, let one now deal with the **main theme** of the present chapter as expressed in the **caption** of this chapter.

The absolute or the final truth regarding the **concrete, current cosmos** or the **physical, present cosmos** on one hand and **its source** or **creator** on the other is the following.

The creator, maker, progenitor, god, source or whatever of the **concrete, current cosmos** is an incredible or awe-inspiring and **the only one of its kind** or unique, **dimensionless, bodiless** or **anatomyless,** plus **timeless,** non-physical, non-material, unphysical, or immaterial consciousness, awareness, sentience, mind or 'I' of infinite **intelligence, imagination,** and **emotion.**

Incidentally, the **concrete, current cosmos** is the **cosmos** which is noted, detected, observed, perceived and experienced by all the **dimensionless, embodied** or **anatomied** plus **timeless,** non-physical, non-material, unphysical, or immaterial, **human** consciousnesses, awarenesses, sentiences, minds, or 'I's, in their **fully-awake** or **wide-awake-state.**

The above description of the creator, maker, progenitor, god, source or whatever of the **concrete, current cosmos** of mankind's **fully-awake** or **wide-awake-state** has included this creator's, god's or whatever, all the 7 (seven) fundamental attributes, namely that of being **consciousness, dimensionless, bodiless, timeless,** and possession on its part of infinite **intelligence, imagination** and **emotion.**

The above listed 7 (seven) fundamental attributes of the creator, maker, progenitor, god, source or whatever of the **concrete, current cosmos** are like the **Seven Colors of Rainbow** or the **Seven Colors which constitute the Unitary Entity called Sun Light.**

The Seven Colors of Rainbow or the Seven Colors which constitute the Unitary Entity called Sun Light represent a Unitary Phenomenon which is a very apt Symbol of the Universal Idea called the Unity in Diversity or the Nuclear Idea Called Adwaita of the Vedantic Cosmology.

The creator, maker, progenitor, god, source or whatever of the **concrete current cosmos** has created, made, formed, or produced the **concrete, current cosmos** of mankind's **fully-awake** or **wide-awake-state,** by a very common-'o'-garden or ordinary activity on its part called the activity of **daydreaming** or **oneirism** i.e. the activity of **mental-imagery-making** or **consciousnessbal-imagery-making** inside its **mind** or **consciousness,** nothing more nothing less.

Therefore, the **3-D** or **three-dimensional** plus **time-bound,** physical, material, objective, or **concrete, current cosmos** of mankind's **fully-awake** or **wide-awake-state,** is nothing but a mere **daydream** or **oneiric** i.e. is nothing but a mere **mental-imagery** or **consciousnessbal-imagery** of the incredible or awe-inspiring, and **the only one of its kind** or unique,

dimensionless, bodiless or **anatomyless,** plus **timeless,** non-physical, non-material, unphysical, or immaterial, consciousness, awareness, sentience, mind or 'I' of infinite **intelligence, imagination,** and **emotion** called the creator, maker, progenitor, god, source, or whatever of the **concrete, current cosmos,** despite or notwithstanding the absolutely contrary, different, or poles apart plus, unshakable or dogged, belief or faith of the **dimensionless, embodied** or **anatomied** plus **timeless,** non-physical, non-material, unphysical, or immaterial, **human** consciousnesses, awarenesses, sentiences, minds, 'I's of the **concrete, current cosmos** in this matter.

The unshakable or dogged belief or faith of the **dimensionless, embodied** or **anatomied** plus **timeless,** non-physical, non-material, unphysical, or immaterial, **human** consciousnesses, awarenesses, sentiences, minds, 'I's of the **concrete, current cosmos** about the above matter or about the above issue is that the **3-D** or **three-dimensional** plus **time-bound,** physical, material, objective, or **concrete, current cosmos** of mankind's **fully-awake** or **wide-awake-state** is one hundred percent, real, genuine, or authentic plus solid, concrete, objective, physical, material or substantial and in no way, is composed of the **daydream-stuff** or **oneiric-stuff** i.e. and in no way, is composed of **mental-imagery-stuff,** or **consciousnessbal-imagery-stuff** of the **mind** or **consciousness** of its creator, maker, progenitor, god, source or whatever.

And of course they i.e. the **dimensionless, embodied** or **anatomied** plus **timeless,** non-physical, non-material, unphysical, or immaterial, **human** consciousnesses, awarenesses, sentiences, mind's or 'I's of the **concrete, current cosmos** unshakably or doggedly, also believe, maintain, assume or surmise that the creator, maker, progenitor, god, source or whatever of this one hundred percent, real, genuine, or authentic plus solid, objective, physical, material, substantial or **concrete, current cosmos,** lives or abides separate from them in some faraway, mysterious or unknown but stellar, celestial, magnificent, superlative, fantastic, dazzling or divine domain, kingdom or place called heaven, paradise, Shangri-la, Garden of Eden, Swarga, Zion or whatever.

The **3-D** or **three-dimensional** plus **time-bound,** physical, material, substantial, objective, or **concrete, current cosmos** is the **cosmos** which all the **dimensionless, embodied** or **anatomied,** plus **timeless,** and non-physical, non-material, unphysical, or immaterial **human** consciousnesses, awarenesses, sentiences, mind's or 'I's detect, note, observe, perceive and experience, through their five, **3-D** or **three-dimensional, time-bound,** concrete, objective, physical, material, or **substantial sense organs** of their **3-D** or **three-dimensional, time-bound,** concrete, objective, physical, material, or substantial **body** or **anatomy** and in which they all take part through the medium of their **3-D** or **three-dimensional, time-bound,** concrete, objective, physical, material or substantial **body** or **anatomy,** in their **fully-awake** or **wide-awake-state** throughout their existence.

In the light of what has been said above with regards to the **truth** vis-a-vis the **3-D** or **three-dimensional** plus **time-bound,** physical, material, objective or **concrete, current cosmos** of mankind's **fully-awake** or **wide-awake-state** i.e. that this **3-D** or **three-dimensional** plus **time-bound,** physical, material, objective, or **concrete, current cosmos** of mankind's **fully-awake** or **wide-awake-state,** is merely a **daydream** or **oneiric** viz. is merely a **mental-imagery** or

consciousnessbal-imagery of the incredible or awe-inspiring, and **the only one of its kind** or unique, **dimensionless, bodiless** or **anatomyless,** plus **timeless,** non-physical, non-material, unphysical, or immaterial, consciousness, awareness, sentience, mind or 'I' of infinite **intelligence, imagination,** and **emotion** aka the creator, maker, progenitor, god, source, or whatever of the **concrete, current cosmos,** it is now the right moment to point out the **truth** vis-a-vis the unique, ubiquitous, and infinite plus **3-D** or **three-dimensional,** non-physical, non-material, unphysical or immaterial **cosmic space** of mankind's **fully-awake-awake** or **wide-awake-state.**

The **truth** vis-a-vis the unique, ubiquitous, and infinite plus **3-D** or **three-dimensional,** non-physical, non-material, unphysical or immaterial **cosmic space** of mankind's **fully-awake-awake** or **wide-awake-state** is quite different from the **general perception** which the **human** consciousnesses of the **concrete, current cosmos** have with regards to this **cosmic space.**

The **general perception** amongst the **human consciousnesses** with regards to the **cosmic space** of the **concrete current cosmos** is that the **cosmic space** is some kind of **insentient** or incapable of feeling or understanding thing which, by good fortune, came into being from nowhere in the **current cosmos** in order to **spatially** accommodate the existence of the **most important** component of the **current cosmos,** namely, the one hundred percent **physical, material, substantial, objective** or **concrete** component of the **current cosmos,** for example, all the one hundred percent **physical, material, substantial, objective** or **concrete** moons, planets, stars, galaxies and the like of the **current cosmos** on one hand, and all the one hundred percent **physical, material, substantial, objective** or **concrete** bodies or anatomies of all the **embodied** or **anatomied** consciousnesses of the **current cosmos** on the other, for example, all the one hundred percent **physical, material, substantial, objective** or **concrete** bodies or anatomies of all the **embodied** or **anatomied** human consciousnesses of the **current cosmos.**

However, the **truth** with regards to the **cosmic space** of mankind's **fully-awake** or **wide-awake-state** is quite different.

The **truth** with regards to the **cosmic space** of mankind's **fully-awake** or **wide-awake-state** is that this **3-D** or **three-dimensional** plus **timeless,** non-physical, non-material, unphysical or immaterial **cosmic space** is a unique, ubiquitous and infinite, **field of consciousness, awareness, sentience, mind** or **'I'** or, is a unique, ubiquitous and infinite, **ocean of consciousness, awareness, sentience, mind** or **'I',** which has been brought into being by the incredible or awe-inspiring and **the only one of its kind** or unique, **dimensionless, bodiless** or **anatomyless,** plus **timeless,** non-physical, non-material, unphysical or immaterial, consciousness, awareness, sentience, mind or 'I' of infinite **intelligence, imagination,** and **emotion** aka the creator, maker, progenitor, god, source or whatever of the **3-D** or **three-dimensional** plus **time-bound,** physical, material, objective or **concrete current cosmos** of mankind's **fully-awake** or **wide-awake-state.**

The **3-D** or **three-dimensional** plus **timeless,** non-physical, non-material, unphysical or immaterial **current cosmic space** of mankind's **fully-awake** or **wide-awake-state,** is seen or visualized by the **dimensionless, embodied** or

anatomied plus **timeless,** non-physical, non-material, unphysical or immaterial **human** consciousnesses through the medium of the **3-D** or **three-dimensional** plus **time-bound,** physical, material, objective or concrete **eyes** of their **3-D** or **three-dimensional** plus **time-bound,** physical, material, objective or concrete **body** or **anatomy** in their fully-awake or wide-awake-state.

This **3-D** or **three-dimensional** plus **timeless** and non-physical, non-material, unphysical or immaterial, unique, ubiquitous and infinite **cosmic space** of mankind's fully-awake or wide-awake-state or, this **3-D** or **three-dimensional** plus **timeless,** non-physical, non-material, unphysical or immaterial, unique, ubiquitous and infinite **field of consciousness, awareness, sentience, mind** or **'I'** or, this **3-D** or **three-dimensional** plus **timeless,** and non-physical, non-material, unphysical or immaterial, unique, ubiquitous and infinite **ocean of consciousness, awareness, sentience, mind** or **'I'** of mankind's fully-awake or wide-awake-state, which is ignorantly, nesciently, naively or illiterately called the **cosmic space** by the **dimensionless** but **embodied** or **anatomied,** plus **timeless** and non-physical, non-material, unphysical, or immaterial **human** consciousnesses of the **concrete current cosmos** is, in fact, the **expanded, distended, dilated** or **inflated** form or version of the incredible or awe-inspiring and **the only one of its kind** or unique, **dimensionless** and **bodiless** or **anatomyless** plus **timeless,** non-physical, non-material, unphysical or immaterial, consciousness, awareness, sentience, mind or 'I' of infinite **intelligence, imagination,** and **emotion** aka the creator, maker, progenitor, god, source or whatever of the **concrete current cosmos** of mankind's fully-awake or wide-awake-state.

To reiterate.

The incredible or awe-inspiring and **the only one of its kind** or unique, **dimensionless** and **bodiless** or **anatomyless** plus **timeless,** non-physical, non-material, unphysical or immaterial, consciousness, awareness, sentience, mind or 'I' of infinite **intelligence, imagination,** and **emotion** who has become the current or the present-day, **3-D** or **three-dimensional** plus **timeless** and non-physical, non-material, unphysical or immaterial, unique, ubiquitous and infinite **cosmic space** of mankind's fully-awake or wide-awake-state or, the **3-D** or **three-dimensional** plus **timeless,** non-physical, non-material, unphysical or immaterial, unique, ubiquitous and infinite **field of consciousness, awareness, sentience, mind** or **'I'** or, the **3-D** or **three-dimensional** plus **timeless,** and non-physical, non-material, unphysical or immaterial, unique, ubiquitous and infinite **ocean of consciousness, awareness, sentience, mind** or **'I'** of mankind's fully-awake or wide-awake-state through the process of **expansion, distention, dilation** of **inflation** of itself or through the process of **expansion, distention, dilation** of **inflation** of its **dimensionless** and **bodiless** or **anatomyless** plus **timeless,** non-physical, non-material, unphysical or immaterial, consciousness, awareness, sentience, mind, or 'I' of infinite **intelligence, imagination,** and **emotion,** is the incredible or awe-inspiring and **the only one of its kind** or unique god, creator, maker, progenitor, source or whatever of the current or the present-day, **3-D** or **three-dimensional** and **time-bound,** physical, material, objective or **concrete cosmos** of mankind's **fully-awake** or **wide-awake-state.**

The **dimensionless** form or version is the **original** or the **basic** form or version of the **bodiless, timeless,** non-physical, non-material, unphysical or immaterial, consciousness, awareness, sentience, mind, or 'I' of infinite **intelligence, imagination,** and **emotion** aka the creator, maker, progenitor, god, source or whatever of the current **3-D** or **three-dimensional** plus **time-bound,** physical, material, objective or **concrete current cosmos** of mankind's **fully-awake** or **wide-awake-state.** This form or version i.e. the **dimensionless** or the **original** form or version, also known as the **primal** or **basic** form or version of the creator, maker, progenitor, god, source or whatever of the **3-D** or **three-dimensional** plus **time-bound,** physical, material, objective or **concrete current cosmos** of mankind's **fully-awake** or **wide-awake-state** existed some 13.7 billion light years ago only and does not exist now.

The **dimensionless** form or version or, the **original, basic,** or **primal** form or version of the **bodiless** or **anatomyless** plus **timeless,** non-physical, non-material, unphysical or immaterial, consciousness, awareness, sentience, mind or 'I' of infinite **intelligence, imagination,** and **emotion** aka the creator, maker, progenitor, god, source or whatever of the **3-D** or **three-dimensional** plus **time-bound,** physical, material, objective or **concrete current cosmos** of mankind's **fully-awake** or **wide-awake-state,** now, currently, or at present moment, exists as the **3-D** or **three-dimensional** plus **timeless,** non-physical, non-material, unphysical or immaterial, unique, ubiquitous and infinite, **field of consciousness, awareness, sentience, mind** or **'I'** of infinite **intelligence, imagination,** and **emotion,** or, as the **3-D** or **three-dimensional** plus **timeless,** non-physical, non-material, unphysical or immaterial, unique, ubiquitous and infinite, **ocean of consciousness, awareness, sentience, mind** or **'I'** of infinite **intelligence, imagination,** and **emotion** which is ignorantly, nesciently, naively or illiterately called the

cosmic space by the **dimensionless** but **embodied** or **anatomied,** plus **timeless,** non-physical, non-material, unphysical or immaterial **human** consciousnesses, awarenesses, sentiences, minds, 'I's of the **concrete current cosmos.**

As hinted above, the designation or the epithet of **cosmic space** has been pinned onto the **expanded, distended, dilated,** or **inflated** form or version of the creator, maker, progenitor, god or whatever of the **concrete current cosmos,** on account of the lack of the **true knowledge** on the part of the **human** consciousnesses of the **concrete current cosmos** about or regarding the underlying **real nature** of the current **cosmic space** of their **fully-awake** or **wide-awake-state.**

Inside this **3-D** or **three-dimensional** plus **timeless,** non-physical, non-material, unphysical or immaterial, unique, ubiquitous and infinite, **field of consciousness, awareness, sentience, mind** or **'I'** of limitless **intelligence, imagination,** and **emotion** or, the **ocean of consciousness, awareness, sentience, mind** or **'I'** of limitless **intelligence, imagination,** and **emotion** aka **cosmic space** aka god, creator, maker, progenitor, source or whatever of the **concrete current cosmos,** the **3-D** or **three-dimensional, time-bound,** physical, material, objective or **concrete current cosmos** of mankind's **fully-awake** or **wide-awake-state,** is floating, wafting or levitating plus whirling, twirling or spiraling as a mere **daydream** or **oneiric** or, as a mere **mental-imagery** or **consciousnessbal-imagery** of this **3-D** or **three-dimensional** plus **timeless,** non-physical, non-material, unphysical or immaterial, unique, ubiquitous and infinite, **field of consciousness, awareness, sentience, mind** or **'I'** of limitless **intelligence, imagination,** and **emotion** or, the

ocean of consciousness, awareness, sentience, mind or **'I'** of limitless **intelligence, imagination,** and **emotion** aka **cosmic space** aka god, creator maker, progenitor, source or whatever of the **concrete current cosmos** of mankind's **fully-awake** or **wide-awake-state.**

The 3-D or three-dimensional plus timeless, non-physical, non-material, unphysical or immaterial, unique, ubiquitous and infinite, field of consciousness, awareness, sentience, mind or 'I' of limitless intelligence, imagination, and emotion or, the ocean of consciousness, awareness, sentience, mind or 'I' of limitless intelligence, imagination, and emotion of mankind's fully-awake or wide-awake-state, is called or labeled as cosmic space by the dimensionless, embodied or anatomied plus timeless, non-physical, non-material, unphysical or immaterial human consciousnesses, awarenesses, sentiences, minds or 'I's of the concrete current cosmos on account of their ignorance, nescience or naivety or, on account of the lack of the true knowledge on their part about or regarding the underlying real nature of the current cosmic space of their fully-awake or wide-awake-state.

To repeat.

The 3-D or three-dimensional plus timeless, non-physical, non-material, unphysical or immaterial, unique, ubiquitous and infinite, field of consciousness, awareness, sentience, mind or 'I' of limitless intelligence, imagination, and emotion or, the ocean of consciousness, awareness, sentience, mind or 'I' of limitless intelligence, imagination, and emotion is called or labeled as cosmic space by the dimensionless but embodied or anatomied, plus timeless, non-physical, non-material, unphysical or immaterial human consciousnesses, awarenesses, sentiences, minds or 'I's of the concrete current cosmos on account of their ignorance, nescience or naivety or, on account of the lack of the true knowledge on their part about or regarding the underlying real nature of the current cosmic space of their fully-awake or wide-awake-state.

What has been said above can be put in another way.

The current, **3-D** or **three-dimensional** but **time-bound,** physical, material, objective or **concrete current cosmos** of mankind's **fully-awake** or **wide-awake-state,** is floating, wafting or levitating plus whirling, twirling or spiraling non-stop inside the current, **3-D** or **three-dimensional** but **timeless,** non-physical, non-material, unphysical, or immaterial, unique, ubiquitous and infinite **cosmic space** and has been doing so from the beginning of the current time and will continue to do so till the end of the current time, as a mere **daydream** or **oneiric** or, as a mere **mental-imagery** or **consciousnessbal-imagery** of this **3-D** or **three-dimensional** but **timeless,** non-physical, non-material, unphysical or immaterial, unique, ubiquitous and infinite **cosmic space.** In term of **absolute truth,** this **3-D** or **three-dimensional** but **timeless,** non-physical, non-material, unphysical or immaterial, unique, ubiquitous and infinite **cosmic space** is a **3-D** or **three-dimensional** but **timeless,** non-physical, non-material, unphysical or immaterial, unique, ubiquitous and infinite **field of consciousness, awareness, sentience, mind** or **'I'** of limitless **intelligence, imagination,** and **emotion** or, an **ocean of consciousness, awareness, sentience, mind** or **'I'** of limitless **intelligence, imagination,** and **emotion.**

This 3-D or three-dimensional but timeless, non-physical, non-material, unphysical or immaterial, unique, ubiquitous and infinite field of consciousness, awareness, sentience, mind or 'I' of limitless intelligence, imagination, and emotion or, the ocean of consciousness, awareness, sentience, mind or 'I' of limitless intelligence, imagination, and emotion is called or labeled as cosmic space by the dimensionless but embodied or anatomied plus timeless, non-physical, non-material, unphysical or immaterial human consciousnesses, awarenesses, sentiences, minds, or 'I's of the concrete current cosmos, on account of the lack of the true knowledge on their part about or regarding the underlying real nature of the current, 3-D or three-dimensional but timeless, non-physical, non-material, unphysical, or immaterial cosmic space of their fully-awake or wide-awke-state.

The **3-D** or **three-dimensional** but **time-bound,** physical, material, objective or **concrete current cosmos** of mankind's **fully-awake** or **wide-awake-state** consists of countless, one hundred percent, insentient or incapable of feeling or understanding, concrete, objective, physical or **material moons,** planets, stars, galaxies and the like on one hand, and countless, one hundred percent, insentient or incapable of feeling or understanding, concrete, objective, physical, or **material bodies,** or **anatomies** of all the **dimensionless** but **embodied** or **anatomied,** plus **timeless,** non-physical, non-material, unphysical or immaterial consciousnesses, awarenesses, sentiences, minds, or 'I's of the **concrete current cosmos** on the other, for example, all the **dimensionless** but **embodied** or **anatomied,** plus **timeless,** non-physical, non-material, unphysical or immaterial **human** consciousnesses of the physical, material, objective or **concrete current cosmos.**

To sum up.

The unique, ubiquitous, and infinite, **3-D** or **three-**

dimensional plus **timeless,** non-physical, non-material, unphysical or immaterial **cosmic space** of mankind's **fully-awake** or **wide-awake-state** is a unique, ubiquitous, and infinite, **3-D** or **three-dimensional** plus **timeless,** non-physical, non-material, unphysical or immaterial, **field of consciousness** or **ocean of consciousness** of limitless **intelligence, imagination,** and **emotion.**

The above statement is making known the **absolute truth** about or regarding the **current cosmic space** of mankind's **fully-awake-awake** or **wide-awake-state** because, contrary to the popular belief, the **current cosmic space** of mankind's **fully-awake** or **wide-awake-state** is not an enigmatic or inscrutable plus incapable of feeling or understanding, insensate or insentient thing or stuff. On the contrary or contradictorily, it is an incredible or awe-inspiring and **the only one of its kind** or unique, **expanded, distended, dilated** or **inflated** form or version, that is to say, the **3-D** or **three-dimensional** form or version of the **dimensionless** form or version or the **original, primal** or **basic** form or version of the non-physical, non-material, unphysical or immaterial, consciousness, awareness, sentience, mind, or 'I' of the creator, maker, progenitor, god, source or whatever of the **concrete current cosmos** of mankind's **fully-awake** or **wide-awake-state,** nothing more nothing less.

This non-physical, non-material, unphysical or immaterial, consciousness, awareness, sentience, mind, or 'I' of the creator, maker, progenitor, god, source or whatever of the **concrete current cosmos** of mankind's **fully-awake** or **wide-awake-state** possesses infinite **intelligence, imagination,** and **emotion,** irrespective of whether it is in its current **3-D** or **three-dimensional** form or version called **cosmic space**, or it is in its **dimensionless** form or version called its **original,**

primal or **basic** form or version,

To sum up.

The **cosmic space,** seen or viewed by the **embodied** or **anatomied** human consciousnesses in their **fully-awake** or **wide-awake-state,** stands for, represents or symbolizes the **expanded, distended, dilated,** or **inflated** form or version, that is to say, stands for, represents or symbolizes the **3-D** or **three-dimensional** form or version of the **dimensionless** form or version or the **original, basic,** or **primal** form or version of the non-physical, non-material, unphysical or immaterial, consciousness, awareness, sentience, mind, or 'I' of the creator, maker, progenitor, god, source or whatever of the **concrete current cosmos** of mankind's **fully-awake** or **wide-awake-state**

Therefore, the **current cosmic space;** who stands for, represents or symbolizes the **expanded, distended, dilated,** or **inflated** form or version, that is to say, who stands for, represents or symbolizes the **3-D** or **three-dimensional** form or version of the **dimensionless** form or version or the **original, basic,** or **primal** form or version of the non-physical, non-material, unphysical or immaterial, consciousness, awareness, sentience, mind, or 'I' of the creator, maker, progenitor, god, source or whatever of the **concrete current cosmos** of mankind's **fully-awake** or **wide-awake-state;** is a unique, ubiquitous, and infinite, **3-D** or **three-dimensional** plus **timeless,** non-physical, non-material, unphysical or immaterial, **field of consciousness** or **ocean of consciousness** of limitless **intelligence, imagination,** and **emotion.**

The full description of the **dimensionless** form or version or the **original, primal** or **basic** form or version of the non-physical, non-material, unphysical or immaterial, consciousness, awareness, sentience, mind or 'I' of the creator, maker, progenitor, god, source or what ever of the **3-D** or **three-dimensional** plus **time-bound,** physical, material, objective or **concrete current cosmos** of mankind's **fully-awake** or **wide-awake-state** is the following.

" It is the **bodiless** or **anatomyless** plus **dimensionless** & **timeless,** non-physical, non-material, unphysical or immaterial, consciousness, awareness, sentience, mind or 'I' of limitless or infinite **intelligence, imagination,** and **emotion".** Therefore, in terms of **absolute truth,** the **current cosmic space** of mankind's **fully-awake** or **wide-awake-state** is not an enigmatic or inscrutable plus incapable of feeling or understanding, insentient or insensate thing or stuff. Instead, it is an incredible or awe-inspiring and **the only one of the kind** or unique, **expanded, distended, dilated** or **inflated** form or version, that is to say, the **3-D** or **three-dimensional** form or version of the **dimensionless** form or version or the **original, primal** or **basic** form or version of the **bodiless** or **anatomyless** plus **timeless,** non-physical, non-material, unphysical or immaterial, consciousness, awareness, sentience, mind or 'I' of infinite **intelligence, imagination,** and **emotion** of god, creator, maker, progenitor, source or whatever of the **concrete current cosmos** of mankind's **fully-awake** or **wide-awake-state,** nothing more nothing less.

Just to remind oneself that this incredible or awe-inspiring and **the only one of its kind** or unique, creator, maker, progenitor, god, source or whatever of the **concrete current cosmos** of mankind's **fully-awake** or **wide-awake-state** is **bodiless** or **anatomyless** plus **timeless,** non-physical, non-material, unphysical or immaterial, consciousness, awareness, sentience, mind or 'I' of infinite **intelligence, imagination,** and **emotion.** It is intrinsically capable of existing in two forms or versions namely, the **dimensionless** form or version on one hand, and the **3-D** or **three-dimensional** form or version on the other. It does not exist in both forms or versions at one and at the same time. At any given point, it exists either in one form or the other. At the present moment, it exists in its **3-D** or **three-dimensional** form or version which is called or labeled as **cosmic space** by **human** consciousnesses on account of the lack on their part of the **true knowledge** about or regarding the **real nature** of the **current cosmic space.**

Hence, what is called or labeled as **cosmic space** by the **human** consciousnesses or awarenesses in their **fully-awake** or **wide-awake-state**, in terms of **absolute reality** or **fact,** is the **expanded, distended, dilated,** or **inflated** form or version or, the **3-D** or **three-dimensional** form or version of the **dimensionless** form or version or the **original, primal** or **basic** form or version of god's or creator's consciousness or awareness. This incredible or awe-inspiring and **the only one of its kind** or unique **consciousness** or **awareness** of god, creator, or whatever of the **concrete current cosmos** of mankind's **fully-awake** or **wide-awake-state** possesses limitless or infinite **intelligence, imagination,** and **emotion** from the perspective of the current **human** consciousnesses or awarenesses.

The **dimensionless, bodiless** or **anatomyless** plus **timeless** form or version of the non-physical, non-material, unphysical or immaterial, consciousness, awareness, sentience, mind or 'I' of the creator, maker, progenitor, god, source or whatever of the **3-D** or **three-dimensional** plus **time-bound,** physical, material, objective or **concrete**

current cosmos of mankind's **fully-awake** or **wide-awake-state** is also called its **primal** form or version, in addition to being called its **original** or **basic** form or version.

The above described **original, primal,** or **basic** form or version of the non-physical, non-material, unphysical or immaterial, consciousness, awareness, sentience, mind or 'I' of the creator, maker, progenitor, god, source, or whatever of the **3-D** or **three-dimensional** plus **time-bound,** physical, material, objective or **concrete current cosmos** of mankind's **fully-awake** or **wide-awake-state** exists only when it is not **daydreaming** or only when it is not involved in **oneirism.**

That is to say, the above described, **original, primal,** or **basic** form or version of the non-physical, non-material, unphysical or immaterial, consciousness, awareness, sentience, mind or 'I' of the creator, maker, progenitor, god, source, or whatever of the **3-D** or **three-dimensional** plus **time-bound,** physical, material, objective or **concrete current cosmos** of mankind's **fully-awake** or **wide-awake-state** exists only when it is not involved in **mental-imagery-making** or **consciousnessbal-imagery-making** inside itself.

In other words, the above described, **original, primal,** or **basic** form or version of the non-physical, non-material, unphysical or immaterial, consciousness, awareness, sentience, mind or 'I' of the creator, maker, progenitor, god, source, or whatever of the **3-D** or **three-dimensional** plus **time-bound,** physical, material, objective or **concrete current cosmos** of mankind's **fully-awake** or **wide-awake-state** exists only when it is not involved in **mental-imagery-making** or **consciousnessbal-imagery-making** inside its **original, primal** or **basic** form or version aka **dimensionless** form or version of its **bodiless** or **anotomyless** plus **timeless,** non-physical, non-material, unphysical or immaterial consciousness, awareness, sentience, mind or 'I'.

The **original, primal,** or **basic** form or version of the **bodiless** or **anatomyless** plus **timeless,** non-physical, non-material, unphysical or immaterial, consciousness, awareness, sentience, mind or 'I' of the creator, maker, progenitor, god, source, or whatever of the **3-D** or **three-dimensional** plus **time-bound,** physical, material, objective or **concrete current cosmos** of mankind's **fully-awake** or **wide-awake-state,** existed some 13.7 billion light years ago and does not exist now. It exists now as the current **3-D** or **three-dimensional,** non-physical, non-material, unphysical or immaterial **cosmic space** or, as the current **3-D** or **three-dimensional,** non-physical, non-material, unphysical or immaterial, unique, ubiquitous and infinite, **field of consciousness, awareness, sentience, mind** or **'I'** of limitless **intelligence, imagination,** and **emotion** or, as the current **3-D** or **three-dimensional,** non-physical, non-material, unphysical or immaterial, unique, ubiquitous and infinite, **ocean of consciousness, awareness, sentience, mind** or **'I'** of limitless **intelligence, imagination,** and **emotion** which is called or labeled, as **cosmic space** by the **dimensionless, embodied** or **anatomied,** plus **timeless,** non-physical, non-material, unphysical or immaterial **human** consciousnesses of the **concrete current cosmos,** on account of the lack on their part of the true **knowledge** about the underlying **real nature** of the **current cosmic space** of their fully-awake or wide-awake-state.

To put it simply.

The **3-D** or **three-dimensional** plus **time-bound,** physical, material, objective, or **concrete current cosmos** which all the

dimensionless, embodied or **anatomied** plus **timeless,** non-physical, non-material, unphysical or immaterial **human** consciousnesses of the **concrete current cosmos** detect, note, observe, perceive and experience plus take part in their **fully-awake** or **wide-awake-state,** is a mere **daydream** or **oneiric** i.e. is a mere **mental imagery** or **consciousnessbal imagery** of the **3-D** or **three-dimensional** plus **timeless,** non-physical, non-material, unphysical, or immaterial **current cosmic space**, despite or notwithstanding the absolutely contrary, different, or poles apart plus unshakable or dogged belief or faith of the **dimensionless, embodied** or **anatomied** plus **timeless,** non-physical, non-material, unphysical, or immaterial, **human** consciousnesses of the **concrete current cosmos** in this matter.

The unshakable or dogged belief or faith of the **dimensionless, embodied** or **anatomied** plus **timeless,** non-physical, non-material, unphysical, or immaterial, **human** consciousnesses of the **current concrete cosmos** about the above matter or about the above issue is that the **3-D** or **three-dimensional** plus **time-bound,** physical, material, objective, or **concrete current cosmos** of mankind's **fully-awake** or **wide-awake-state** is one hundred percent, real, genuine, or authentic plus solid, concrete, objective, physical, material or substantial and in no way, it is composed of the **daydream-stuff** or **oneiric-stuff** i.e. and in no way, it is composed of **mental-imagery-stuff** or **consciousnessbal-imagery-stuff** of its creator, maker, progenitor, god, source or whatever.

And of course they i.e. the **dimensionless, embodied** or **anatomied** plus **timeless,** non-physical, non-material, unphysical, or immaterial, **human** consciousnesses, awarenesses, sentiences, minds or 'I's of the **concrete current cosmos,** unshakably or doggedly, also believe, maintain, assume or surmise that the creator, maker, progenitor, god, source or whatever of this one hundred percent, real, genuine, or authentic plus solid, physical, material, substantial, objective or **concrete current cosmos,** lives or abides separate from them in some unknown but fantastic place called heaven, paradise, Shangri-la, Garden of Eden, Swarga, Zion or whatever.

Just to remind oneself that the **3-D** or **three-dimensional** plus **time-bound,** physical, material, objective, or **concrete current cosmos** of mankind's fully-awake or wide-awake-state consists of countless, insentient or incapable of feeling or understanding, objective, concrete, physical or **material moons,** planets, stars, galaxies and the like on one hand and countless, insentient or incapable of feeling or understanding, physical, material, objective or concrete bodies or anatomies of all the countless, **embodied** consciousnesses of the **current cosmos** on the other, for example, all the insentient or incapable of feeling or understanding, physical, material, objective or concrete, bodies or anatomies of all the **embodied** human consciousnesses of the **current cosmos.**

And in the end, let one remind oneself that the **3-D** or **three-dimensional** plus **timeless,** non-physical, non-material, unphysical or immaterial, unique, ubiquitous and infinite **current cosmic space** of mankind's fully-awake or wide-awake-state is a unique, ubiquitous and infinite, **3-D** or **three-dimensional** plus **timeless,** non-physical, non-material, unphysical or immaterial **field of consciousness, awareness, sentience, mind** or 'I' of limitless **intelligence, imagination,** and **emotion** or, is a unique, ubiquitous and infinite, **3-D** or **three-dimensional** plus **timeless,** non-physical, non-material, unphysical or immaterial, **ocean of consciousness, awareness, sentience, mind** or 'I' of limitless **intelligence, imagination,** and **emotion** and it is in fact the creator, maker, progenitor, god, source, or whatever of the **3-D** or **three-dimensional** plus **time-bound,** insentient or incapable of feeling or understanding, physical, material, objective or **concrete current cosmos.**

The **3-D** or **three-dimensional** plus **timeless,** non-physical, non-material, unphysical or immaterial, unique, ubiquitous and infinite **current cosmic space** of mankind's fully-awake or wide-awake-state is called or labeled as **cosmic space** by the human consciousnesses on account of the lack on their part of the **true knowledge** about the underlying **real nature** of this **3-D** or **three-dimensional** plus **timeless,** non-physical, non-material, unphysical or immaterial, unique, ubiquitous and infinite **current cosmic space** of their fully-awake or wide-awake-state.

THE REASON WHY HUMAN CONSCIOUSNESSES ARE THE WAY THEY ARE

The **human** consciousnesses of the **concrete-current-cosmos** or the **objective-current-cosmos** are described as follows.

They, viz. **human** consciousnesses of the **concrete-current-cosmos** or the **objective-current-cosmos** are the non-physical, non-material, unphysical, or unsubstantial, or, not composed of matter, **dimensionless,** plus **timeless, deathless, immortal** or **eternal** but physically, materially, substantially, or objectively, **embodied** or **anatomied,** consciousnesses.

The above description of **human** consciousnesses of the **concrete-current-cosmos** or the **objective-current-cosmos** lists the 5 (five) fundamental attributes of **human**

consciousnesses, namely :-
They are consciousness.
They are not composed of matter.
They are dimensionless.
They are timeless, deathless, immortal or eternal.
They are physically, materially, substantially, or objectively, embodied or anatomied. Their physical, material, substantial or objective body or anatomy alone which is time-bound, death-bound, mortal, perishable, impermanent or ephemeral. They themselves are, namely human consciousnesses themselves are timeless, deathless, immortal or eternal in the manner of their source, god or whatever, namely the cosmic space of mankind's fully-awake or wide-awake-state.

When one says that **human** consciousnesses are **physically, materially, substantially,** or **objectively, embodied** or **anatomied,** one means that **human** consciousnesses are not **bodiless** or **anatomyless** in the manner of their **source, god** or whatever, namely the **immaterial-and-3-D-or-three-dimensional-current-cosmic-space** of mankind's fully-awake or wide-awake-state who, which or whatever is the source, god or whatever of all the **embodied** or **anatomied** consciousnesses of the **concrete-current-cosmos** or the **objective-current-cosmos,** on one hand, for example, all the **human** consciousnesses, and the creator, maker, progenitor or whatever of the **concrete-current-cosmos** or the **objective-current-cosmos**, on the other.

The immaterial-and-3-D-or-three-dimensional-current-cosmic-space of mankind's fully-awake or wide-awake-state; who, which or whatever is the source, god or whatever of all the **embodied** or **anatomied** consciousnesses of the **concrete-current-cosmos** or the **objective-current-cosmos,** on one hand, for example, all the **human**

consciousnesses, and the creator, maker, progenitor or whatever of the **concrete-current-cosmos** or the **objective-current-cosmos**, on the other; is a **seeable-truth, visible-truth, discernible-truth, manifest-truth, revealed-truth, out-in-the-open-truth, big-as-life-truth, under-one's-nose-truth** or **not-hidden-truth,** despite being non-physical, non-material, unphysical or unsubstantial because it is ocularly seen, viewed, discerned, observed or visualised by all the **embodied** or **anatomied** human consciousnesses of the **concrete-current-cosmos** or the **objective-current-cosmos** through the medium of the **concrete** or the **objective** eyes of their **concrete** or the **objective** body or anatomy.
At this juncture it will be cogent, relevant or logical to recount the complete list of all the fundamental attributes of the **immaterial-and-3-D-or-three-dimensional-current-cosmic-space** of mankind's fully-awake or wide-awake-state aka the source, god or whatever of all the **embodied** or **anatomied** consciousnesses of the **concrete-current-cosmos** or the **objective-current-cosmos,** on one hand, for example, all the **human** consciousnesses, and the creator, maker, progenitor or whatever of the **concrete-current-cosmos** or the **objective-current-cosmos**, on the other.

Let one specify here and now that at this juncture it will be cogent, relevant or logical to recount the complete list of all the fundamental attributes of the **original** or the **primal** form or version of the **immaterial-and-3-D-or-three-dimensional-current-cosmic-space** of mankind's fully-awake or wide-awake-state aka the source, god or whatever of all the **embodied** or **anatomied** consciousnesses of the **concrete-current-cosmos** or the **objective-current-cosmos,** on one hand, for example, all the **human** consciousnesses, and the creator, maker, progenitor or whatever of the **concrete-current-cosmos** or the **objective-current-cosmos**, on the other, meaning thereby, when the latter, namely the **immaterial-and-3-D-or-three-dimensional-current-cosmic-space** was in its **original** or **primal** form or version which existed some 13.7 billion light years ago and does not exist now.

To recount the complete list of all the fundamental attributes of the **original** or the **primal** form or version of the **immaterial-and-3-D-or-three-dimensional-current-cosmic-space** of mankind's fully-awake or wide-awake-state is cogent, relevant or logical at this juncture because this **immaterial-and-3-D-or-three-dimensional-current-cosmic-space** of mankind's fully-awake or wide-awake-state, is bound to revert or return to its **original** or **primal** form or version sometime in the future when it becomes completely bored, fatigued, or sick and tired of watching its present **daydream, oneiric, mental-imagery** or **consciousnessbal-imagery,** or, if it is preferred, when it becomes completely bored, fatigued, or sick and tired of watching its present **phantasmagoria, shifting series of phantasms, shifting series of phenomena, shifting series of illusions,** or **shifting series of deceptive appearances** called, labeled or designated as the **concrete-current-cosmos** or the **objective-current-cosmos** of mankind's fully-awake or wide-awake-state which it, namely the **immaterial-and-3-D-or-three-dimensional-current-cosmic-space** of mankind's fully-awake or wide-awake-state has given rise to by a common-'o'-garden or very ordinary activity on its part, called the activity of **daydreaming, oneirism, mental-imagery-making** or **consciousnessbal-imagery-making** and nothing else.

As said above, the **original** or the **primal** form or version of the **immaterial-and-3-D-or-three-dimensional-current-**

cosmic-space of mankind's fully-awake or wide-awake-state aka the source, god or whatever of all the **embodied** or **anotomied** consciousnesses of the **concrete-current-cosmos** or the **objective-current-cosmos,** on one hand, for example, all the **human** consciousnesses, and the creator, maker, progenitor or whatever of the **concrete-current-cosmos** or the **objective-current-cosmos**, on the other, existed some 13.7 billion light years ago and does not exist now. It now exists as the **immaterial-and-3-D-or-three-dimensional-current-cosmic-space** of mankind's fully-awake or wide-awake-state.

The immaterial-and-3-D-or-three-dimensional-current-cosmic-space of mankind's fully-awake or wide-awake-state aka the source, god or whatever of all the embodied or anotomied consciousnesses of the concrete-current-cosmos or the objective-current-cosmos, on one hand, for example, all the human consciousnesses, and the creator, maker or progenitor of the concrete-current-cosmos or the objective-current-cosmos, on the other, possesses 7 (seven) fundamental attributes when it is in its original or primal form or version, namely that of being an incredible or awe-inspiring plus the only one of its kind or unique, bodiless and dimensionless consciousness who or which or whatever is also timeless, deathless, immortal or eternal on one hand, and possesses infinite intelligence, imagination and emotion on the other.

To sum up.

The immaterial-and-3-D-or-three-dimensional-current-cosmic-space of mankind's fully-awake or wide-awake-state aka the source, god or whatever of all the embodied or anatomied consciousnesses of the concrete-current-cosmos or the objective-current-cosmos, on one hand, for example, all the human consciousnesses, and the creator, maker or progenitor of the concrete-current-cosmos or the objective-current-cosmos, on the other, possesses the following 7 (seven) fundamental attributes, when it is in its original or primal form or version :-

It is consciousness.
It is dimensionless.
It is bodiless.
It is timeless, deathless, immortal or eternal.
It possesses infinite **intelligence.**
It possesses infinite **imagination.**
It possesses infinite **emotion.**

It will be quite appropriate at this point to place together for contrasting effect all the 7 (seven) fundamental attributes of the **original** or **primal** form or version of the **immaterial-and-3-D-or-three-dimensional-current-cosmic-space** of mankind's fully-awake or wide-awake-state and all the 5 (five) fundamental attributes of **human** consciousnesses of the **concrete-current-cosmos** or **objective-current-cosmos.**

Here are the 5 (five) fundamental attributes of **human** consciousnesses.

They are consciousness.
They are not composed of matter.
They are dimensionless.
They are timeless, deathless, immortal or eternal.
They are physically, materially, substantially, or objectively, embodied or anatomied. Their physical, material, substantial or objective body or anatomy alone which is time-bound, death-bound, mortal, perishable, impermanent or ephemeral.

They themselves are, that is to say, human consciousnesses themselves are timeless, deathless, immortal or eternal in the manner of their source, god or whatever, namely the cosmic space of mankind's fully-awake or wide-awake-state.

Now let us deal with the answer to the question posed in the caption or the heading of this chapter, namely **"why human consciousnesses are the way they are"?**

The reason why the non-physical, non-material, unphysical, or unsubstantial, or, not composed of matter, **dimensionless,** plus **timeless, deathless, immortal** or **eternal** but physically, materially, substantially, or objectively, **embodied** or **anatomied, human** consciousnesses of the **concrete-current-cosmos** or the **objective-current-cosmos** are **the way they are,** that is to say, the reason why the non-physical, non-material, unphysical, or unsubstantial, or, not composed of matter, **dimensionless,** plus **timeless, deathless, immortal** or **eternal** but physically, materially, substantially, or objectively, **embodied** or **anatomied, human** consciousnesses of the **concrete-current-cosmos** or the **objective-current-cosmos** have an unshakable or dogged belief or faith that the **3-D** or **three-dimensional** plus **time-bound, death-bound, destined-to-die, short-lived** or **impermanent** physical, material, objective or **concrete-current-cosmos** or the **objective-current-cosmos** is one hundred percent, real, genuine, or authentic plus solid, concrete, objective, physical, material or substantial and, therefore in no way, is composed of the **daydream-stuff** or **oneiric-stuff** i.e. and, therefore in no way, is composed of the **mental-imagery-stuff** or **consciousnessbal-imagery-stuff** of its creator, maker, progenitor or whatever, namely the **immaterial-and-3-D-or-three-dimensional-current-cosmic-space** of mankind's fully-awake or wide-awake-state, is the following.

Firstly, all the physical, material, objective, or concrete **bodies** or **anatomies** of all the non-physical, non-material, unphysical, or unsubstantial, or, not composed of matter, **dimensionless,** plus **timeless, deathless, immortal** or **eternal** but physically, materially, substantially, or objectively, **embodied** or **anatomied, human** consciousnesses of the **concrete-current-cosmos** or the **objective-current-cosmos,** are nothing but one amongst the countless other physical, material, objective or concrete **things, items** or **component parts** of the **3-D** or **three-dimensional** plus **time-bound, death-bound, destined-to-die, short-lived** or **impermanent** physical, material, objective or **concrete-current-cosmos** or the **objective-current cosmos.**

Secondly, the non-physical, non-material, unphysical, or unsubstantial, or, not composed of matter, **dimensionless,** plus **timeless, deathless, immortal** or **eternal** but physically, materially, substantially, or objectively, **embodied** or **anatomied, human** consciousnesses of the **concrete-current-cosmos** or the **objective-current-cosmos** have a deep but absolutely erroneous belief that their **"true identity"** or **"the fact of their true being, who or what they truly are",** consists of their **3-D** or **three-dimensional** plus **time-bound, death-bound, destined-to-die** or, better still, **destined-for-death-and-destruction,** physical, material, objective or concrete **body** or **anatomy** only and nothing else.

As a consequence of their deep even though absolutely erroneous belief that their "true identity" or "the fact of their true being, who or what they truly are" comprises or consists of their 3-D or three-dimensional plus time-bound, death-bound, destined-to-die or destined-for-death-and-destruction, physical, material, objective or concrete body or anatomy only and nothing else, the human consciousnesses of the concrete-current-cosmos or the objective-current-cosmos develop or form an intense or fervent opinion inside their minds or consciousnesses that they themselves are time-bound, death-bound, destined-to-die or destined-for-death-and-destruction one day, in the manner of their physical, material, objective or concrete body or anatomy.

The deep albeit absolutely erroneous belief of the human consciousnesses of the concrete-current-cosmos or the objective-current-cosmos that their "true identity" or "the fact of their true being, who or what they truly are" comprises or consists of their 3-D or three-dimensional plus time-bound, death-bound, destined-to-die or destined-for-death-and-destruction, physical, material, objective or concrete body or anatomy only and nothing else, sinks them neck-deep or almost to the point of total submersion into the very limited and transient ocean of the physical, material, substantial, objective or concrete current cosmos.

And since in the **very limited** and **transient ocean** of physical, material, substantial, objective or **concrete-current-cosmos** or **objective-current-cosmos,** every, thing or item, is physical, material, substantial, objective or concrete plus transient, temporary, ephemeral, time-bound, death-bound, short-lived or impermanent and nothing else, **human** consciousnesses are deprived of any independent **yard-stick,** measure, scale, model, gauge, benchmark, indicator, criterion, norm, basis, standard, sample, or example to judge, assess, work-out, find out, or figure out or, if it is preferred, solve, settle, see through, or size up or, better still, accomplish, establish, or clinch **what is real** and **what is unreal** or **what is true** and **what is false** or, to be extremely exact, **what is eternally real** and **what is** only **temporarily real,** unless and until someday they decide to look beyond the **very limited** and **transient ocean** of physical, material, substantial or **concrete-current-cosmos** or **objective-current-cosmos** and begin to look towards the **cosmic space** and towards **themselves** namely the human **consciousnesses** in order seek out that which is **limitless** and **deathless** or **limitless** and **timeless,** namely the **cosmic space,** on one hand, and human **consciousnesses** on the other.

All that has been said above can be put in another way.

The non-physical, non-material, unphysical, or unsubstantial, or, not composed of matter, **dimensionless,** plus **timeless, deathless, immortal** or **eternal** but physically, materially, substantially, or objectively, **embodied** or **anatomied** human consciousnesses of the **concrete-current-cosmos** or the **objective-current-cosmos** have an unshakable or dogged belief or faith that the **3-D** or **three-dimensional** plus **time-bound, death-bound, destined-to-die** or **destined-for-death-and-destruction,** physical, material, objective or **concrete-current-cosmos** or **objective-current-cosmos** of mankind's **fully-awake** or **wide-awake-state** is one hundred percent, real, genuine, or authentic on one hand and solid, concrete, objective, physical, material or substantial on the other.

All the **3-D** or **three-dimensional** plus **time-bound, death-bound, destined-to-die** or **destined-for-death-and-destruction,** physical, material, substantial, objective, or concrete **bodies** or **anatomies** of all the non-physical, non-material, unphysical, or unsubstantial, or, not composed of matter, **dimensionless,** plus **timeless, deathless, immortal** or **eternal** human consciousnesses, are in fact made of **daydream-stuff** or **oneiric-stuff** i.e. are in fact made of **mental-imagery-stuff** or **consciousnessbal-imagery-stuff** of the **immaterial-and-3-D-or-three-dimensional-current-cosmic-space** aka the creator, maker, progenitor or whatever of the **3-D** or **three-dimensional** plus **time-bound, death-bound, destined-to-die** or **destined-for-death-and-destruction, concrete-current-cosmos** or **objective-current-cosmos** of mankind's **fully-awake** or **wide-awake-state.**

When the immaterial-and-3-D-or-three-dimensional-current-cosmic-space aka the creator, maker, progenitor or whatever of the 3-D or three-dimensional plus time-bound, death-bound, destined-to-die or destined-for-death-and-destruction, concrete-current-cosmos or objective-current-cosmos of mankind's fully-awake or wide-awake-state is in its original or primal form or version, it exists as an incredible or awe-inspiring plus the only one of its kind or unique, non-physical, non-material, unphysical, or unsubstantial, dimensionless and bodiless plus timeless, deathless, immortal or eternal consciousness, awareness, sentience, mind, or 'I' of infinite intelligence, imagination, and emotion.

However, today the immaterial-and-3-D-or-three-dimensional-current-cosmic-space aka the creator, maker, progenitor or whatever of the 3-D or three-dimensional plus time-bound, death-bound, destined-to-die or destined-for-death-and-destruction, concrete-current-cosmos or objective-current-cosmos of mankind's fully-awake or wide-awake-state does not exist in its original or primal form or version.

In other words, today the immaterial-and-3-D-or-three-dimensional-current-cosmic-space aka the creator, maker, progenitor or whatever of the 3-D or three-dimensional plus time-bound, death-bound, destined-to-die or destined-for-death-and-destruction, concrete-current-cosmos or objective-current-cosmos of mankind's fully-awake or wide-awake-state does not exist in its dimensionless form or version. Instead, it exists today, in its expanded, distended, dilated, or inflated form or version or, better still, instead, it exists today in the expanded, distended, dilated, or inflated form or version of its dimensionless form or version or original form or version.

The **expanded, distended, dilated,** or **inflated** form or version of the **dimensionless** form or version or **original** form or version of the creator, maker, progenitor or whatever of the **concrete-current-cosmos** or the **objective-current-cosmos** of mankind's fully-awake or wide-awake-state has been given the name, appellation or designation of **cosmic space** by human consciousnesses of the **concrete-current-cosmos** or **objective-current-cosmos.**

The 3-D or three-dimensional **cosmic space** of today, (just like its predecessor, namely its **dimensionless** form or version or **original** form or version), is the incredible or awe-inspiring plus **the only one of its kind** or unique, non-physical, non-material, unphysical, or unsubstantial, **timeless, deathless, immortal** or **eternal,** consciousness, awareness, sentience, mind, or 'I', of infinite **intelligence, imagination,** and **emotion.** Or to be absolutely exact, the 3-D or three-dimensional **cosmic space** of today, (just like its predecessor, namely its **dimensionless** form or version or **original** form or version), is the incredible or awe-inspiring plus **the only one of its kind** or unique, non-physical, non-material, unphysical, or unsubstantial, **timeless, deathless, immortal** or **eternal** consciousness, awareness, sentience, mind, or 'I' of infinite **intelligence, imagination,** and **emotion** of the creator, maker, progenitor or whatever of the **concrete current cosmos** or **objective-current-cosmos** of mankind's fully-awake or wide-awake-state.

In other words, contrary to the popular belief prevalent amongst the **human** consciousnesses of today, the 3-D or three-dimensional plus non-physical, non-material, unphysical or unsubstantial, **cosmic space** is not an insentient or incapable of feeling or understanding thing or entity. Instead, it is an incredible or awe-inspiring plus **the only one of its kind** or unique, non-physical, non-material, unphysical, or unsubstantial, **timeless, deathless, immortal** or **eternal** consciousness, awareness, sentience, mind, or 'I' of infinite **intelligence, imagination,** and **emotion** of the creator, maker, progenitor or whatever of the **concrete-current-cosmos** or **objective-current-cosmos** of mankind's fully-awake or wide-awake-state, nothing more nothing less.

And inside this **3-D** or **three-dimensional,** incredible or awe-inspiring plus **the only one of its kind** or unique, non-physical, non-material, unphysical, or unsubstantial, **timeless, deathless, immortal** or **eternal** consciousness, awareness, sentience, mind, or 'I' of infinite **intelligence, imagination,** and **emotion** called **cosmic space** by human beings or, absolutely to the point, inside this **expanded, distended, dilated,** or **inflated** form or version of the originally **dimensionless** immortal creator, maker, progenitor or whatever of infinite **intelligence, imagination,** and **emotion** who has been presently given the name of **cosmic space** by human beings, today's **3-D** or **three-dimensional** plus **time-bound, death-bound, destined-to-die** or **destined-for-death-and-destruction,** concrete-current-cosmos or objective-current-cosmos of mankind's **fully-awake** or **wide-awake-state** is floating, wafting, or levitating plus whirling, twirling, or spiralling non-stop as a mere **daydream** or **oneiric** or, as a mere **mental-imagery** or **consciousnessbal-imagery** of this incredible immortal creator, maker, progenitor or whatever of infinite **intelligence, imagination,** and **emotion** who is called **cosmic space** by human beings of today.

To repeat.

And inside this **3-D** or **three-dimensional,** incredible or awe-inspiring plus **the only one of its kind** or unique, non-physical, non-material, unphysical, or unsubstantial, **timeless, deathless, immortal** or **eternal** consciousness, awareness, sentience, mind, or 'I' of infinite **intelligence, imagination,** and **emotion** called **cosmic space** by human beings or, absolutely to the point, and inside this **expanded, distended, dilated,** or **inflated** form or version of the originally **dimensionless** immortal creator, maker, progenitor or whatever of infinite **intelligence, imagination,** and **emotion** who has been presently given the name of **cosmic space** by human beings, today's **3-D** or **three-dimensional** plus **time-bound, death-bound, destined-to-die** or **destined-for-death-and-destruction,** concrete-current-cosmos or objective-current-cosmos of mankind's **fully-awake** or **wide-awake-state** is floating, wafting, or levitating plus whirling, twirling, or spiralling non-stop as a mere **daydream** or **oneiric** or, as a mere **mental-imagery** or **consciousnessbal-imagery** of this incredible immortal creator, maker, progenitor or whatever of infinite **intelligence, imagination,** and **emotion** who is called **cosmic space** by human beings of today.

Today's 3-D or three-dimensional plus time-bound, death-bound, destined-to-die, or destined-for-death-and-destruction, **concrete-current-cosmos** or **objective-current-cosmos** of mankind's **fully-awake** or **wide-awake-state** has been floating, wafting, or levitating plus whirling, twirling, or spiralling non-stop as a mere **daydream** or **oneiric** or, as a mere **mental-imagery** or **consciousnessbal-imagery** of the present-day **cosmic space** aka the incredible immortal creator, maker, progenitor or whatever of infinite **intelligence, imagination,** and **emotion** from the beginning of the current time and will continue to do so till the end of the current time.

The intrinsic nature of all the 3-D or three-dimensional plus time-bound, death-bound, destined-to-die or, destined-for-death-and-destruction, physical, material, substantial, objective or concrete, bodies or anatomies, which constitute the bodies or anatomies, of all the non-physical, non-material, unphysical, or unsubstantial, or, not composed of matter, dimensionless, plus timeless, deathless, immortal or eternal but physically, materially, substantially, or objectively, embodied or anatomied, consciousnesses of the concrete-current-cosmos or objective-current-cosmos is exactly the same as that described vis-a-vis all the 3-D or three-dimensional plus time-bound, death-bound, destined-to-die, or destined-for-death-and-destruction, physical, material, substantial, objective, or concrete bodies or anatomies, of all the non-physical, non-material, unphysical, or unsubstantial, or, not composed of matter, dimensionless, plus timeless, deathless, immortal or eternal but physically, materially, substantially, or objectively, embodied or anatomied human consciousnesses of the concrete-current-cosmos or objective-current-cosmos.

That is to say, all the **3-D** or **three-dimensional** plus **time-bound, death-bound, destined-to-die,** or **destined-for-**

death-and-destruction, physical, material, substantial, objective, or concrete **bodies** or **anatomies** of all the other non-physical, non-material, unphysical, or unsubstantial, or, not composed of matter, **dimensionless,** plus **timeless, deathless, immortal** or **eternal** but physically, materially, substantially, or objectively, **embodied** or **anatomied,** consciousnesses of the **concrete-current-cosmos** or **objective-current-cosmos,** for example, all the animal and plant **consciousnesses** of the **concrete-current-cosmos** or **objective-current-cosmos** are also made of nothing but the **daydream-stuff** or **oneiric-stuff** i.e. are also made of nothing but the **mental-imagery-stuff** or **consciousnessbal-imagery-stuff** of the present-day **cosmic space** aka the creator, maker, progenitor or whatever of the **3-D** or **three-dimensional** plus **time-bound, death-bound, destined-to-die,** or **destined-for-death-and-destruction** concrete-current-cosmos or objective-current-cosmos.

Therefore, the present-day **cosmic space** is nothing but the creator, maker, progenitor or whatever of the present-day **3-D** or **three-dimensional** plus **time-bound, death-bound, destined-to-die,** or **destined-for-death-and-destruction** concrete-cosmos or objective-cosmos of mankind's fully-awake or wide-awake-state.

The non-physical, non-material, unphysical, or unsubstantial, or, not composed of matter, **dimensionless,** plus **timeless, deathless, immortal** or **eternal** but physically, materially, substantially, or objectively, **embodied** or **anatomied** human consciousnesses of the present-day **concrete-cosmos** or **objective-cosmos** develop an unshakable or dogged belief or faith from their very early life that their "**true identity**" or "**the fact of their true being who or what they truly are**", consists of their **3-D** or **three-dimensional** plus **time-bound,**

death-bound, destined-to-die, or **destined-for-death-and-destruction,** physical, material, substantial, objective or concrete **body** or **anatomy** only and nothing else.

That is to say, the non-physical, non-material, unphysical, or unsubstantial, or, not composed of matter, **dimensionless,** plus **timeless, deathless, immortal** or **eternal** but physically, materially, substantially, or objectively, **embodied** or **anatomied,** human consciousnesses of the present-day **concrete-cosmos** or **objective-cosmos** develop an unshakable or dogged belief or faith from their very early life that **they are nothing more than their 3-D** or **three-dimensional** plus **time-bound, death-bound, destined-to-die** or **destined-for-death-and-destruction** physical, material, objective or concrete **body** or **anatomy.**

This is indeed very unfortunate that the non-physical, non-material, unphysical, or unsubstantial, or, not composed of matter, **dimensionless,** plus **timeless, deathless, immortal** or **eternal** but physically, materially, substantially, or objectively, **embodied** or **anatomied** human consciousnesses of the present-day **concrete-cosmos** or **objective-cosmos** develop an unshakable or dogged belief or faith from their very early life that their **"true identity"** or **"the fact of their true being who or what they truly are",** consists of nothing **more than their 3-D** or **three-dimensional** plus **time-bound, death-bound, destined-to-die,** or **destined-for-death-and-destruction,** physical, material, objective or concrete **body** or **anatomy.**

The "iron grip or should one say, the "death grip" of the "absolutely false idea" on the timeless, deathless, immortal or eternal human consciousnesses of the concrete-current-cosmos or objective-current-cosmos that their "true identity" or "the fact of their true being who or what they truly are", consists of nothing more than their 3-D or three-dimensional plus time-bound, death-bound, destined-to-die, or destined-for-death-and-destruction physical, material, objective or concrete body or anatomy is so strong, powerful, vigorous, fierce, intense, or extreme that most of them remain under the "iron grip" or, should one say, the "death grip" of this "absolutely false idea" every moment of their entire existence in the concrete-current-cosmos or objective-current-cosmos. They are never able to "free themselves" of this "iron grip or, should one say, "death grip" of this "absolutely false idea".

What has been said above can be put in another way.

The "iron grip or should one say, the "death grip" of the "absolutely false idea" on the timeless, deathless, immortal or eternal human consciousnesses of the concrete-current-cosmos or objective-current-cosmos that they are nothing but only their 3-D or three-dimensional plus time-bound, death-bound, destined-to-die, or destined-for-death-and-destruction physical, material, objective or concrete body or anatomy is so strong, powerful, vigorous, fierce, intense, or extreme that most of them remain under the "iron grip" or, should one say, the "death grip" of this "absolutely false idea" every moment of their entire existence in the concrete-current-cosmos or objective-current-cosmos. They are never able to "free themselves" of this "iron grip or, should one say, "death grip" of this "absolutely false idea".

The above described "absolutely false idea" is called the "cosmic nescience", "cosmic obtuseness", "cosmic denseness", "cosmic blindness" or, in Sanskrit, the "brahmandic avidya", "brahmandic agyana", "brahmandic jadta", or "brahmandic moorkhta".

The **absolute truth** is that the **human** consciousnesses are the non-physical, non-material, unphysical, or unsubstantial, or, not composed of matter, **dimensionless,** plus **timeless, deathless, immortal** or **eternal,** but physically, materially, substantially, or objectively, **embodied** or **anatomied** consciousnesses.

In other words, the **human** consciousnesses are not the **3-D** or **three-dimensional** plus **time-bound, death-bound, destined-to-die** or **destined-for-death-and-destruction** physical, material, objective or concrete **body** or **anatomy** under any condition or circumstance.

The 3-D or three-dimensional plus time-bound, death-bound, destined-to-die, or destined for death and destruction, physical, material, objective or concrete body or anatomy of human consciousnesses is merely a temporary dwelling, abode or place of residence for the human consciousnesses during the period of their brief, fleeting, or short sojourn or stopover in the concrete-current-cosmos or objective-current-cosmos, nothing more nothing less.

The human **consciousnesses** themselves are **timeless, deathless, immortal** or **eternal** in the manner of their source or god or whatever, namely **cosmic space** of the **concrete current cosmos** or **objective current cosmos** of mankind's fully-awake or wide-awake-state because they all are nothing but an absolutely pure or pristine section, segment, part, of portion of **cosmic space** aka the creator, maker, progenitor or whatever of the **concrete current cosmos** or **objective current cosmos.**

To sum up.

In terms of truth, the concrete-current-cosmos or objective-current-cosmos of mankind's fully-awake or wide-awake-state, is a phantasmagoria, a web of illusion or a shifting series of appearances on account of the fact that it is a daydream, oneiric, mental-imagery or consciousnessbal-imagery of the incredible cosmic space aka the creator, maker, progenitor or whatever of the concrete-current-cosmos or objective-current-cosmos of mankind's fully-awake or wide-awake-state.

The phantasmagoria, web of illusion, shifting series of appearances, daydream, oneiric, mental-imagery or consciousnessbal-imagery called or labeled as the concrete-current-cosmos or objective-current-cosmos of mankind's fully-awake or wide-awake-state which has been produced or formed by cosmic space aka the creator, maker, progenitor or whatever through a common-'o'-garden or very ordinary activity called daydreaming or oneirism on its part, is given the name, appellation or designation of "maya-jaal" in the Adwait-Vedantic Realm because it is a very apt or appropriate name, appellation or designation for it.

The name or appellation **"maya-jaal"** for the **concrete-current-cosmos** or **objective-current-cosmos** of mankind's fully-awake or wide-awake-state is very apt or appropriate because this name or appellation tells what's the nature of the **concrete-current-cosmos** or **objective-current-cosmos** of mankind's fully-awake or wide-awake-state truly is, notwithstanding or despite the totally contrarian or antagonistic viewpoint of mankind at large on this subject, topic or issue who regard the **concrete current cosmos** or **objective-current-cosmos** of their fully-awake or wide-awake-state to be absolutely genuine, authentic or real and made or composed of solid or concrete **physical matter,** the solid or concrete **physical matter** which can not only be seen by their physical eyes but can also be smelt by their physical nose, tasted by their tongue, heard by their physical ears, touched by their physical fingers and stamped on by their physical feet.

The meaning of the Sanskrit word **"maya-jaal"** is the following :- a **shifting series of phantasms,** a **shifting series of illusions,** or a **shifting series of deceptive appearances** as in a dream or as created by someone's imagination or flights of fancy.

As said before, the "maya-jaal", phantasmagoria, web of illusion, shifting series of deceptive appearances or shifting series of phantasms called or labeled as concrete-current-cosmos or the objective-current-cosmos of mankind's fully-awake or wide-awake-state, has been formed or produced by cosmic space aka the creator, maker, progenitor or whatever through a very ordinary or common-'o'-garden activity on its part called the activity of daydreaming or oneirism or, the activity of mental-imagery-making or consciousnessbal-imagery-making, nothing more nothing les

FUNDAMENTALLY THERE IS ONLY ONE THING IN THE CURRENT COSMOS AND THAT THING IS COSMIC SPACE - 1

As the title of this chapter unambiguously states, there is only one thing in the current cosmos of **fundamental importance** and that thing is the **cosmic space.** This is so because rest of the items of the current cosmos, for example, **physical matter** and its twin **physical energy,** not forgetting all the embodied or anatomied **consciousnesses** of the current cosmos, for example, all the human **consciousnesses,** have been given rise to or have been brought into existence in the current cosmos by this **"big daddy of all"** of the current cosmos namely the **cosmic space** of mankind's fully-awake or wide-awake-state.

So, the question naturally arises inside human **minds** or human **consciousnesses** who or what this **"big daddy of all"** of the current cosmos, namely the current **cosmic space** of mankind's fully-awake or wide-awake-state **truly** is in terms of its intrinsic **nature** and how it itself has come into **being** in the current cosmos on one hand, and what is its **role** or **function** in the current cosmos on the other.

It has already been stated earlier that this **"big daddy of all"** of the current cosmos i.e. the current **cosmic space** of mankind's fully-awake or wide-awake-state is the **fountainhead** or the **originator** of all the items of the current cosmos namely the **physical matter/physical energy duo** on one hand, and all the embodied or anatomied **consciousnesses** of the current cosmos on the other, for example, all the human **consciousnesses.** But the question still remains how and why this **"big daddy of all"** of the current cosmos, namely the **cosmic space** has managed to give rise to the amazing **physical universe** inside itself, not forgetting, countless embodied or anatomied **consciousnesses** of the current cosmos, for example, human **consciousnesses,** inside itself.

This **"big daddy of all"** of the current cosmos namely the **cosmic space** is an incredible or awe-inspiring plus **the only one of its kind** or unique, non-physical, non-material, unphysical or unsubstantial, ubiquitous and infinite, 3-D or three-dimensional, **field of consciousness,** awareness, sentience, mind or 'I' of unbounded intelligence, imagination and emotion.

As said before, rest of items which are presently existent in the current cosmos and are thus, perceived and experienced by the human **consciousnesses** of the current cosmos in their fully-awake or wide-awake-state, have all emanated from this incredible **field of consciousness,** namely the current **cosmic space** i.e. the **"big daddy of all"** of the current cosmos. Even the human **consciousnesses** themselves who perceive and experience all the **physical** items of the current cosmos have emanated from this incredible **field of consciousness,** namely the **cosmic space** who, which or whatever is here being repeatedly referred to as the **"big daddy of all"** of the current cosmos.

This **"big daddy of all"** of the current cosmos, namely the **cosmic space** is not an insentient or incapable of feeling or understanding thing or entity. Instead, as said before, it is an incredible **field of consciousness** who is one hundred per aware, sensate or sentient and therefore, animated, spirited, lively, sprightly, dynamic and ebullient, not in a common--'o'-garden, human-way, but in its own inscrutable, **cosmic-way.** It possesses infinite **intelligence, imagination** & **emotion.**

Of course, this **"big daddy of all"** of the current cosmos, namely the **cosmic space** is uniquely **bodiless,** meaning thereby it does not don or wear a personal **body** or **anatomy** in the manner all human **consciousnesses** do. That is to say, it does not personally don or wear a body or anatomy in the manner, all human **consciousnesses** have been made to don or wear a body or anatomy by this **"big daddy of all"** of the current cosmos i.e. the **cosmic space** for his, her, its, or whatever **necessity** or **compulsion** which was to give rise to or bring about variety, diversity, multiplicity or variegation inside its featureless or nondescript, **expanded, distended, dilated** or **inflated** consciousness aka **cosmic space** aka the **"big daddy of all"** of the current cosmos during its common-o'-garden or very ordinary activity called **daydreaming** or **oneirism** in order to amuse, entertain or regale itself, nothing more nothing less.

However, the consequence of being **bodiless** or

anatomyless for this **"big daddy of all"** of the current cosmos, is that it does not possess a **physical** speech-equipment in the manner **physically** embodied human **consciousnesses** possess. This speech equipment of human **consciousnesses** consists of the **physical** lips and lungs plus larynx and tongue of their **body** or **anatomy.**

The possession of physical **body** or **anatomy** by the human **consciousnesses** and its associated physical **speech equipment** has enabled the human **consciousnesses** to verbally or physically communicate with each other.

Since the incredible **field of consciousness** of the current cosmos, namely the **cosmic space** aka the **"big daddy of all"** of the current cosmos does not possess a physical **speech equipment** i.e. lips and lungs plus larynx and tongue due to the lack of **physical** body or **anatomy** on its part, it is not possible for him, her, it or whatever to **verbally** or **physically** communicate or speak to human **consciousnesses** of the current cosmos. As a consequence of this, the human **consciousnesses** of the current cosmos have universally concluded that this **"big daddy of all"** of the current cosmos, namely the **cosmic space;** the **cosmic space** which they all see or visualise through the medium of the eyes of their **physical** body or anatomy; is an insentient or incapable of feeling or understanding thing or entity. This has indeed proved very unfortunate for the human **consciousnesses.**

This very unfortunate conclusion arrived at universally by all the human **consciousnesses** of the current cosmos that the **"big daddy of all"** of the current cosmos, namely the **cosmic space** is an insentient or incapable of feeling or understanding thing or entity, has made some human

consciousnesses of the current cosmos to accept **"the absolute supremacy of the physical matter"** in the current cosmos and also prompted them either to overlook completely or to downplay the fact of existence of their own **consciousnesses** on one hand and the fact of existence of the **"big daddy of all"** of the current cosmos, namely the **cosmic space** on the other.

Some amongst this group of human **consciousnesses** have gone on to the extent of becoming extremely aggressive, unrestrained and absolute **materialists.** Their **materialistic approach** has become so extreme that they pronounce unambiguously that all the **consciousnesses** which are extant in the current cosmos, including all the human **consciousnesses,** are merely a **byproduct** of the **physical matter** of the current cosmos.

Even those human **consciousnesses** who possess the inkling, idea or notion or, more to the point, faith, belief, confidence or conviction that there is more to the current cosmos than **physical matter** or, better still, even those human **consciousnesses** who possess the inkling, idea, notion, faith, belief, confidence or conviction that there is something or someone much bigger or higher than **physical matter** who stands behind the current cosmos in order to support, underpin or buttress it, opine, conjecture, theorise or hypothesise that this something or someone, who is above and beyond the **physical matter** and stands behind the current cosmos in order to support, underpin or buttress it, exists outside the current cosmos in some mysterious place called **the heaven, the paradise, the promised land, the City of God, the celestial city, the abode of God, the Zion, the empyrean, the next world, the afterworld, the Swarga** or whatever.

These believers assert the existence of **a single and absolute truth** behind the current cosmos but they opine that this single and absolute truth transcends the current cosmos and is not immanent in the current cosmos. That is to say, they opine that this single and absolute truth which is standing behind the current cosmos, does not pervade or permeate the current cosmos. In other words, it is quite aloof from the affairs of the current cosmos and merely acts as an all -powerful ruler, judge, jury, and executioner and also the rewarder vis-a-vis all human beings of the current cosmos. As per the viewpoint of the human **consciousnesses** of these faiths, this unique and indivisible, all powerful, single and absolute being is independent of its creation namely the current cosmos. These human **consciousnesses** reject the **binary modes of thinking** such as the idea of the **existence of duality** in this single and absolute monarch of the current cosmos or in this single and absolute truth of the current cosmos which stands behind the current cosmos in order to support, underpin or buttress it.

That is to say, as per the viewpoint of the human **consciousnesses** of these faiths, this unique and indivisible, all powerful, single and absolute being is **independent of its creation** namely the current cosmos. These human **consciousnesses** reject the **binary modes of thinking** which put forward the concept of **immanence** in the current cosmos of this single and absolute truth meaning thereby that both **good** and **evil,** which are profusely existent in the current cosmos, emanate, generate or flow from this single and absolute truth's **creative act** only i.e. **daydreaming** or **oneirism** only and nothing else.

FUNDAMENTALLY THERE IS ONLY ONE THING IN THE CURRENT COSMOS AND THAT THING IS COSMIC SPACE - 2

Let us explore further the reason why the incredible or awe-inspiring and **the only one of its kind** or unique, non-physical, non-material, unphysical or unsubstantial, ubiquitous and infinite, 3-D or three-dimensional **cosmic space,** which is seen by all the human **consciousnesses** of the current cosmos in their fully-awake or wide-awake-state through the medium of the **physical eyes** of their 3-D or three-dimensional **physical body** or **anatomy** and inside which all the **physical** moons, planets, stars, galaxies and the like on one hand, and all the physical **bodies** or **anatomies** of all the **embodied** or **anatomied** consciousnesses of the current **physical cosmos** on the other, for example, all the physical **bodies** or **anatomies** of all the **human** consciousnesses, are floating, wafting, or levitating plus whirling, twirling or spiralling non-stop, has been given the mantle, badge or crown of being **the most important** or **significant** truth of the current cosmos.

First of all, let one reiterate that the **cosmic space** which is seen by all the human **consciousnesses** of the current cosmos through the medium of the **physical** eyes of their **physical** body or anatomy in their fully-awake or wide-awake-state, is not an insentient or incapable of feeling or understanding thing or entity as opined by the human **consciousnesses** of the current cosmos. Instead, it is an extraordinary **field of consciousness** who, which or whatever, by virtue of being a **field of consciousness,** is also spirited, peppy, zippy, lively, sprightly, dynamic or ebullient, or, if it preferred, is also alert, awake, bustling, eager, energetic, vigorous or zestful. To top it all, it also is replete with or overflowing with boundless **intelligence, imagination** and **emotion.**

This ubiquitous and infinite 3-D or three-dimensional **field of consciousness** aka **cosmic space** aka the "**big daddy of all**" of the current cosmos, inside which all the **physical** moons, planets, stars, galaxies and the like on one hand, and all the physical **bodies** or **anatomies** of all the **embodied** or **anatomied** consciousnesses of the current **physical cosmos** on the other, for example, all the physical **bodies** or **anatomies** of all the human **consciousnesses,** are floating, wafting, or levitating plus whirling, twirling or spiralling non-stop, is an **immortal, imperishable, indestructible, everlasting, eternal, timeless, deathless, endless** or **never-ending** truth.

However, one must realise that this **immortal, imperishable, indestructible, everlasting, eternal, timeless, deathless, endless** or **never-ending** truth, namely the **cosmic space** of mankind's fully-awake or wide-awake-state does not always or forever exist in its 3-D or three-dimensional form, version, mode, state, or condition.

Cosmic space aka the **"big daddy of all"** of the current cosmos of mankind's fully-awake or wide-awake-state exists in its 3-D or three-dimensional form, version, mode, state or condition only when it is engaged or busy in its common-'o'-garden or very ordinary activity called **daydreaming** or **oneirism** or, if it is preferred, only when it is engaged or busy in its common-'o'-garden or very ordinary activity called **mental-imagery-making** or **consciousnessbal-imagery-making** inside itself, that is to say, inside its **mind** or **consciousness** in order to form or give rise to a **daydream-stuff, oneiric-stuff, mental-imagery-stuff** or **consciousnessbal-imagery-stuff** cosmos, universe or world inside its **mind** or **consciousness** in order to amuse, entertain or regale itself, nothing more nothing less. Otherwise this extraordinary **cosmic space** aka the **"big daddy of all"** of the current cosmos exists in its **dimensionless** form, version, mode, state or condition.

In other words, when this extraordinary **cosmic space** aka the **"big daddy of all"** of the current cosmos is not engaged or busy in its common-'o'-garden or very ordinary activity called **daydreaming, oneirism, mental-imagery-making** or **consciousnessbal-imagery-making** inside itself, that is to say, inside its **mind** or **consciousness** in order to form or give rise to a **daydream-stuff, oneiric-stuff, mental-imagery-stuff** or **consciousnessbal-imagery-stuff** cosmos, universe or world inside its **mind** or **consciousness** in order to amuse, entertain or regale itself, it exists in its **dimensionless** form, version, mode, state or condition.

Either way or irrespective of whether this extraordinary **cosmic space** aka the **"big daddy of all"** of the current cosmos exists in its 3-D or three-dimensional form, version, mode, state or condition, as is the case at present, or, in its **dimensionless** form, version, mode, state or condition, which was the case some 13.7 billion light years ago, it always is and it always will be an amazing, incredible or extraordinary **immortal, imperishable, indestructible, everlasting, eternal, timeless, deathless, endless** or **never-ending** plus spirited, peppy, zippy, lively, sprightly, dynamic or ebullient, or, if it preferred, alert, awake, bustling, eager, energetic, vigorous or zestful **consciousness, awarenesses, sentience, mind** or **'I'** who, which or whatever is also eternally replete, awash or overflowing with boundless **intelligence, imagination** and **emotion.**

This extraordinary **cosmic space** aka the **"big daddy of all"** of the current cosmos has as many names as there are human **consciousnesses** in the current cosmos. Some of these names are listed below :-

God, Brahman, the Creator, the Maker, the Progenitor, the Author etc. etc. of the current cosmos.

The extraordinary, 3-D or three-dimensional **field of consciousness** aka **cosmic space** aka the **"big daddy of all"** of the current cosmos is the **expanded, distended, dilated** or **inflated** form or version of its **dimensions** form or version which existed some 13.7 billion light years ago.

As far as all the **physically, materially, substantially, objectively** or **concretely** embodied or anatomied **consciousnesses** of the current cosmos is concerned, for example, all the human **consciousnesses** of the current cosmos, they all without exception are an absolutely unmodified, unchanged or unaltered plus absolutely pure, pristine, original or virginal segment, section, part or portion of this extraordinary **field of consciousness** called **cosmic space** aka the **"big daddy of all"** of the current cosmos.

Last but not least, all the **physical matter** of the current cosmos which composes all the **physical** moons, planets, stars, galaxies and the like on one hand and all the **physical** bodies or anatomies of all the **physically** embodied or anatomied **consciousnesses** of the current cosmos on the other, for example, the **physical** bodies or anatomies of all the human **consciousnesses,** is made, built or formed of a **condensed, compressed, compacted** or **congealed** section, segment, part or portion of this extraordinary **field of consciousness** called **cosmic space** aka the **"big daddy of all"** of the current cosmos. This extraordinary **field of consciousness** aka **cosmic space** aka the **"big daddy of all"** of the current cosmos, as said before, is also called god, Brahman, creator, maker, progenitor, author or whatever of the current cosmos.

HUMAN CONSCIOUSNESSES! IT IS THE COSMIC SPACE ON WHOM YOU MUST FOCUS YOUR FULL ATTENTION

The caption or the title of this chapter exhorts, urges, counsels or, beckons, appeals or invites the human **consciousnesses** of the current cosmos to concentrate their mind, psyche, nous, wits or intellectual gifts on **cosmic space** of the current cosmos one pointedly if they are seeking, searching, pursuing or looking for the answer vis-a-vis or with regards to their own source on one hand and the source of the current **physical** cosmos on the other.

Dabbling, tinkering or flirting with physical matter alone, as they are doing at present, is no good for human **consciousnesses** if they are searching for the **definitive, unequivocal, conclusive** or **decisive** answer to the above two questions.

Human **consciousnesses,** during their sojourn or stopover in the current **physical** cosmos, spend most of their allotted time in foraging or searching for food, clothing, shelter, entertainment, procreation and sleep. This is very unfortunate indeed.

Whether human **consciousnesses** know or not, the present-day 3-D or three-dimensional **cosmic space** is not only the provider of the **vital space** for the territorial **placement** and **existence** of the current 3-D or the three-dimensional **physical** cosmos but additionally, it forms the pivot, hub or focal point around which the entire being, truth or existence of the current **physical cosmos** gyrates.

The present-day **cosmic space** which is a breathtaking or spectacular 3-D or three-dimensional **field of consciousness,** was formed or created 13.7 billion light years ago, by god, Brahman, the creator, the maker, the progenitor, the source, the author or whatever of the current **physical cosmos,** through the process of **expansion, distension, dilation** or **inflation** of its **dimensionless** form or version or, the **primal** or **original** form or version. This act of **expansion, distension, dilation** or **inflation** of its **dimensionless** form or version or, **primal** or **original** form or version on the part of god, Brahman, the creator, the maker, the progenitor, the source, the author or whatever of the current **physical** cosmos, was due to the exigency or need or, was perforce or of necessity because the **physical** cosmos is a 3-D or three-dimensional object, thing or truth in terms of its configuration, layout, design, shape, form, outline or silhouette and consequently, the **dimensionless** form or version, or, the **primal** or **original** form or version of the **consciousness** or **mind** of god, Brahman, the creator, the maker, the progenitor, the source, the author or whatever of the current **physical** cosmos was in no position to accommodate such a 3-D or three-dimensional entity as current **physical** cosmos.

From what has been said above, it emerges or transpires that if god, Brahman, the source, the creator, the maker, the progenitor, the author or whatever of the current **physical** cosmos desires at any time; during the course of its extraordinary and unique **immortal, eternal, everlasting, never-ending, timeless, deathless** or **endless** existence; to form or give rise to a 3-D or three-dimensional **physical** cosmos inside its **consciousness** or **mind** through the process of **daydreaming** or **oneirism** on its part in order to amuse, entertain or regale itself because it is feeling bored by its own eternal and lone company, it has no option, recourse or choice before him, her, it or whatever other than to first **expand, distend, dilate** or **inflate,** the **dimensionless** form or version of its **consciousness** or **mind** or, the **primal** form or version of its **consciousness** or **mind** or, the **original** form or version of its **consciousness** or **mind** before it can form or give rise to a 3-D or three-dimensional, **daydream-stuff-composed** or **oneiric-stuff-composed,** physical cosmos, universe or world inside its **consciousness** or **mind.**

And this is exactly what it did 13.7 billion light years ago. Ever since then, god, Brahman, the source, the creator, the maker, the progenitor, the author or whatever of the current **physical** cosmos has remained in its **expanded, distended, dilated** or **inflated** form or version & this **expanded, distended, dilated** or **inflated** form or version of god, Brahman, the source, the creator, the maker, the progenitor, the author or whatever of the current **physical** cosmos has been given the name or label of **cosmic space** by the human **consciousnesses** of the present-day **physical** cosmos.

God, Brahman, the creator, the maker, the progenitor, the author or whatever of the current **physical** cosmos will remain in its current form or version namely its **expanded, distended, dilated** or **inflated** form or version; which has been given the name or label of **cosmic space** by human **consciousnesses;** as long as this god, Brahman, the creator, the maker, the progenitor, the author or whatever of the current **physical cosmos** wants to keep in existence some form or the other of its current 3-D or three-dimensional **daydream-stuff-composed** or **oneiric-stuff-composed** physical cosmos inside its **expanded, distended, dilated** or **inflated consciousness** or **mind** i.e. current **cosmic space,** for its amusement, entertainment or regalement.

It is worth emphasising once again that the current **physical** cosmos, which is seen, observed or perceived plus experienced by all human **consciousnesses** of the current **physical** cosmos in their fully-awake or wide-awake-state, has been' formed or given rise to by god, Brahman, the creator , the maker , the progenitor, the author, the source or whatever of the current **physical** cosmos inside its **expanded, distended, dilated** or **inflated** consciousness or mind by a very ordinary or common-'o'-garden process called **daydreaming** or **oneirism** i.e **mental-imagery-making** or **consciousnessbal-imagery-making** on its part, in order to amuse, entertain or regale itself, nothing more nothing less. And this **expanded, distended, dilated** or **inflated** consciousness or mind of god, Brahman, the creator, the maker, the progenitor, the author, the source or whatever of the current **physical** cosmos is none other than the entity called **cosmic space** by human **consciousnesses** of the current physical **cosmos.**

~*~*~*~*~

PHYSICAL COSMOS IS AN OBJECTIVE FACT VIS-A-VIS HUMAN CONSCIOUSNESS BUT A SUBJECTIVE THING VIS-A-VIS ITS CREATOR

Physical cosmos, even though it looks, feels and sounds solid and therefore is nothing but an objective fact vis-a-vis **embodied** human consciousness, rock-bottom truth is that it is nothing of the sort. It is an **oneiric** or **daydream** of god and, thus, is a **subjective** thing as far as god is concerned because this is the absolute reality at the level of god, notwithstanding the **delusion** of **embodied** human consciousness in this regard.

Embodied human consciousness takes **physical** cosmos as being **solid** and **objective** because **embodied** human consciousness is part and parcel of this **physical** cosmos and therefore, neck-deep in it or, therefore, sunk, absorbed or involved almost to the point of total submersion in it. Consequently, **embodied** human consciousness possesses no absolutely independent yardstick, point of reference, benchmark, paradigm or gauge to assess whether **physical** cosmos is really **objective** or fundamentally **subjective** which merely seems **objective** to **embodied** human consciousness due only to the fact that the latter namely **embodied** human consciousness is a part and parcel of this **physical** cosmos and therefore neck-deep in it or, therefore, sunk, absorbed or involved almost to the point of total submersion in it and thus possesses no independent yardstick to assess whether **physical** cosmos is truly **objective** or fundamentally **subjective.**

Embodied human consciousness is part and parcel of **physical** cosmos only because body of **embodied** human consciousness is part and parcel of **physical** cosmos due to the fact that body of **embodied** human consciousness is also an **oneiric** or **daydream** of god just as is the entire **physical matter** of the **physical** cosmos.

However, human consciousness itself is not an **oneiric** or **daydream** of god. Instead, it is a one hundred percent genuine **portion, piece** or **chunk** of god's consciousness and thus, it possesses all the intrinsic pluses and minuses of god's consciousness.

The only difference that obtains between god's **non-embodied** consciousness and **embodied** human consciousness is with regards to their respective **size** and obviously with regards to all the differences that surface between them on account of the difference in their **size** plus due to the fact that one amongst these two is **embodied** i.e. human consciousness and the other is **non-embodied** i.e. god's consciousness.

Human consciousness, who is a one hundred percent genuine **portion, piece** or **chunk** of god's **non-embodied** consciousness, has to be enclosed in a body or better still, has to be clothed with a sheath of body or, if it is preferred, has to be transfigured into an **embodied** consciousness by its source namely god because of the following reason.

For god's innately **non-embodied** consciousness, **embodiment** or, better still, **embodification, personification, manifestation, representation, expression** or **incarnation** of a portion, piece or chunk of its own **non-embodied** consciousness in a tangible or visible form was the one and the only means open to it at the time of its resolve about 13.7 billion light years ago, to bring about **variety, diversity, multiplicity** or **heterogeneity** inside its unremarkable or uninteresting **consciousness** or, if it is preferred, inside its featureless or nondescript **consciousness.**

What has been said above can be put in another way.

Causing **embodification** or **embodiment** of a portion, piece or chunk of its innately **non-embodied** consciousness was the one and the only means open to god's **non-embodied** consciousness at the time of its resolve about 13.7 billion light years ago to give rise to **variety, diversity, heterogeneity** or **multiplicity** inside its **nondescript** or **featureless** consciousness through its activity of **daydreaming** or **oneirism.**

Activity of **daydreaming** or **oneirism,** either undertaken by god's **non-embodied** consciousness or undertaken by man's **embodied** consciousness is also described as **consciousnessbal-imagery-making** or **consciousnessbal-**

dreamry-making.

Irrespective of whether the **daydreamer** or the **oneiricker** in question is the **non-embodied** consciousness of god or the **embodied** consciousness of man, in both cases the **daydream** in question or the **oneiric** in question is situated inside the consciousness of its **daydreamer** or **oneiricker.**

Or, if it is preferred, irrespective of whether the **daydream-maker** or the **oneiric-maker** in question is the **non-embodied** consciousness of god or the **embodied** consciousness of man, in both cases, the **daydream** in question or the **oneiric** in question is situated inside the consciousness of its **daydream-maker** or **oneiric-maker.**

The purpose of saying all that has been said above is to bring to the notice of the **embodied** human consciousness the following **information.**

Just as a **daydream** or **oneiric** created by an **embodied** human
consciousness is situated inside that **embodied** human consciousness only and nowhere else, similarly a **daydream** or **oneiric** created by the **non-embodied** consciousness of god is situated inside the **non-embodied** consciousness of god only and nowhere else.

And the **non-embodied** consciousness of god in the current cosmos is no other entity than the **cosmic space** of the current cosmos.

This **cosmic space** of the current cosmos is seen by the **embodied** human consciousness of the current cosmos in its fully-awake or wide-awake-state via the media of eyes of its

physical body.

And the **daydream** or the **oneiric** of the **non-embodied** consciousness of god is no other entity than the **physical matter** of the current cosmos.

To repeat.

The **non-embodied** consciousness of god is no other entity than the **cosmic space** of the current cosmos and the **daydream** or the **oneiric** of the **non-embodied** consciousness of god is no other entity than the **physical matter** of the current cosmos.

The **physical matter** of the current cosmos is seen by the **embodied** human consciousness in its fully-awake or wide-awake-state via the media of eyes of its **physical** body just as the **cosmic space** of the current cosmos is seen via the media of the eyes of its **physical** body.

What has been said above can be put in a slightly different way.

Cosmic space is the non-embodied, dimensionless consciousness of god in its **expanded, distended, dilated** or **inflated** form or version.

And inside this non-embodied, dimensionless consciousness of god aka **cosmic space,** its **daydream** or **oneiric** namely the **physical** cosmos is floating, wafting or levitating plus whirling, twirling or spiraling non-stop and has been doing so from the beginning of time and will continue to do so till the end of time.

This **daydream** or **oneiric** of non-embodied, dimensionless consciousness of god aka **cosmic space** consists of countless **physical** moons, planets, stars, galaxies and the like on one hand, and countless **physical** bodies of all the **embodied** consciousnesses of the **physical** cosmos on the other, for example, the **physical** bodies of all the **embodied** human consciousnesses of the **physical** cosmos.

To repeat.

Cosmic space is the non-embodied, dimensionless consciousness of god in its **expanded, distended, dilated** or **inflated** form or version and **physical** cosmos is god's i.e. **cosmic space's** daydream-stuff-composed or oneiric-stuff-composed creation.

Thus, all the daydream-stuff-composed or oneiric-stuff-composed **subjective** items of the **daydream** or **oneiric** of this god i.e. cosmic space, are floating, wafting or levitating plus whirling, twirling or spiraling non-stop inside this god's consciousness i.e. inside this cosmic space, and has been doing so from the beginning of time and will continue to do so till the end of time.

All these daydream-stuff-composed or oneiric-stuff-composed **subjective** items of the **daydream** or **oneiric** of god or, absolutely to the point, all these daydream-stuff-composed or oneiric-stuff-composed **subjective** items of the **daydream** or **oneiric** of cosmic space, are called **objective** items i.e. **physical** items by the human consciousness on account of its delusion. One of these **subjective** items vis-a-vis god's consciousness or, absolutely to the point, one of these **subjective** items vis-a-vis cosmic space, is the **body** of human consciousness. On account of its delusion human consciousness calls its **subjective** body an **objective** body or a **physical** body.

There is nothing in the cosmos which is **objective** or **physical** vis-a-vis god's consciousness or, absolutely to the point, there is nothing in the cosmos which is **objective** or **physical** vis-a-vis **cosmic space.**

Everything in the cosmos is **subjective** vis-a-vis god i.e. vis-a-vis **cosmic space.**

All the **subjective** items of the cosmos are accepted as being **objective** i.e. **physical,** by the **embodied** human consciousness on account of the fact that the **embodied** human consciousness is part and parcel of this **subjective** cosmos and therefore, neck-deep in it or, sunk, absorbed or involved almost to the point of total submersion in it.

Consequently, the **embodied** human consciousness possesses no absolutely independent yardstick, point of reference, benchmark, paradigm or gauge to assess whether the **subjective** cosmos is really **objective** or fundamentally **subjective** which merely seems **objective** to the **embodied** human consciousness due only to the fact that the **embodied** human consciousness is part and parcel of this **subjective** cosmos and therefore neck-deep in it or, sunk, absorbed or involved almost to the point of total submersion in it and thus possesses no independent yardstick to assess whether physical cosmos is truly **objective** or fundamentally **subjective.**

THE CREATOR OF THE CURRENT CONCRETE COSMOS IS NOT INVISIBLE. JUST THE REVERSE, IT IS VISIBLE ALL THE TIME TO ALL BECAUSE IT IS THE CURRENT COSMIC SPACE

Embodied human consciousness, on account its delusion, thinks that the creator of the **physical, objective** or **concrete** current cosmos is some **invisible** being who lives in some far -away place called **heaven, paradise, the promised land, the heavenly kingdom, the City of God, the abode of God, the celestial city, Swarg, Elysium, Zion** or whatever and from there it controls the destiny of its creation namely the **physical, objective** or **concrete** current cosmos and also the destiny of all beings living inside it.

Embodied human consciousness is **absolutely** wrong on this issue. The creator of the **concrete** current cosmos does not live in some far -away place called **heaven** or whatever.

Quite the opposite. The creator of the **concrete** current cosmos is the extraordinary entity whom **embodied** human consciousness has given the name **cosmic space** on account of its immense ignorance or, better still, on account of its immense delusion.

Embodied human consciousness sees or visualizes this **cosmic space** via the media of eyes of its **physical, objective** or **concrete** body all the time.

Cosmic space is the consciousness or mind of god or, to be absolutely exact, **cosmic space** is the **consciousnessbal-space** or **mind-space,** situated inside the consciousness or mind of god.

This **consciousnessbal-space** or **mind-space,** situated inside the consciousness or mind of god and which is ignorantly or delusionally called **cosmic apace** by **embodied** human consciousness, has been brought into being or has been formed by god inside itself through the process of **expansion, distention, dilation** or **inflation** of its eternally **dimensionless** or **non-dimensional** consciousness or mind.

God is unlike anything else. It is the only one of its kind or unique, **non-embodied** or **bodiless** consciousness or mind.

All the other consciousnesses or minds of the **physical, objective** or **concrete** current cosmos, for example, all the human and animal consciousnesses or minds of the **physical, objective** or **concrete** current cosmos, are **embodied** or, are in **bodily form** i.e. are **incarnate** or **personified** and not **non-embodied** or **bodiless.**

What has been said above can be put in another way.

God's consciousness or mind is peerless, matchless, unparalleled, or unsurpassed in being **unembodied, disembodied, disbodied, discarnate, incorporeal, ethereal** or in being not existing in **physical, objective** or **concrete** form.

By contrast, human and animal consciousnesses; wherever and whenever they exist or will exist; always are and always will be **embodied, incarnate, personified** or in **bodily form** and never **non-embodied, unembodied, disembodied, disbodied, discarnate** or **bodiless.**

For human and animal consciousnesses or minds to be **non-embodied, unembodied, disembodied, disbodied, discarnate** or **bodiless** in the manner of god's consciousness, is a total impossibility. Why this is a total impossibility in the case of human and animal consciousnesses is a different story and will be addressed or tackled in a different chapter.

God's consciousness or mind is timeless, immortal or eternal. It also is **non-physical, non-material, non-objective** or **non-concrete.** It is eternally **dimensionless** or **non-dimensional.**

When one says that god's consciousness or mind is eternally **dimensionless** or **non-dimensional,** one means that the **dimensionless** or **non-dimensional** nature or character of god's consciousness or mind never changes or alters, irrespective of whether it is in its **expanded, distended, dilated** or **inflated** form or version or, it is in its **un-expanded, un-distended, un-dilated** or **un-inflated** form or version.

God's eternally dimensionless or non-dimensional consciousness or awareness exists in its expanded, distended, dilated or inflated form or version when it daydreams or oneirics or, when it is engaged in its entertainment-time-activity or amusement-time-activity called daydreaming, oneiricking or oneirism. Otherwise it exists in its un-expanded, un-distended, un-dilated or un-inflated form or version.

At the present moment god's eternally **dimensionless** or **non-dimensional** consciousness or mind exists in its **expanded, distended, dilated** or **inflated** form or version because at the present moment it is engaged in **daydreaming, oneiricking** or **oneirism.**

Unfortunately, the current **expanded, distended, dilated** or **inflated** form or version of god's eternally **dimensionless** or **non-dimensional** consciousness or mind has been given the name or epithet of **cosmic space** by the **embodied** human consciousness due to its immense ignorance or delusion.

What has been said above can be put in another way.

God's eternally **dimensionless** or **non-dimensional** consciousness or mind at the present moment exists in its **expanded, distended, dilated** or **inflated** form or version because at the present moment god's eternally **dimensionless** or **non-dimensional** consciousness or mind is engaged in its **entertainment-time-activity** or **amusement-time-activity** called the activity of **daydreaming, oneiricking** or **oneirism.**

The humanly visible evidence or proof of this **entertainment-**

time-activity or **amusement-time-activity** namely, the activity of **daydreaming, oneiricking** or **oneirism,** which is currently in progress inside the eternally **dimensionless** or **non-dimensional** consciousness or mind of god, comprises or consists of the current existence of the entity called **cosmic space** on one hand and the current existence inside this **cosmic space,** of the entity called **physical, objective** or **concrete** cosmos, on the other.

God's consciousness or mind is eternally **dimensionless** or **non-dimensional** in essence or, better still, god's consciousness or mind eternally possesses the attribute of **dimensionless-ness** or **non-dimensional-ness** as the intrinsic and central constituent of its character due to the fact that it is **non-physical, non-material, non-objective** or **non-concrete** and not **physical, objective** or **concrete.**

That is to say, god's consciousness or mind is eternally and quintessentially **dimensionless** or **non-dimensional** due to the fact that it is not composed of **physical, objective** or **concrete matter.**

Physical, objective or concrete matter is always a dimensional thing & never a non-dimensional or dimensionless thing.

In other words, **physical, objective** or **concrete matter;** irrespective of whether it is in its **un-expanded, un-distended, un-dilated** or **un-inflated** form or version or, **expanded, distended, dilated** or **inflated** form or version; always is and always will be a **dimensional** thing and never a **non-dimensional** or **dimensionless** thing. This is the fundamental difference between the consciousness or mind on one hand and the **physical, objective** or **concrete matter** on the other, apart from the fact that the latter is an **insentient** or, incapable of feeling or understanding thing whereas the former is a **sentient** or, capable of feeling or understanding thing.

The role of the **dimensional** and **insentient** physical, objective or concrete matter in the current cosmos is one and one only.

And this one and one only role of the **dimensional** and **insentient** physical, objective or concrete matter in the current cosmos is to become an implement or tool in the hands of god in order to create an array or assemblage of **variety, diversity, heterogeneity** or **multiplicity** in the unremarkable, uninteresting, nondescript or featureless, **expanded, distended, dilated** or **inflated** consciousness or mind of god namely, the current **cosmic space.**

This array or assemblage of **variety, diversity, heterogeneity** or **multiplicity;** inside its unremarkable, uninteresting, nondescript or featureless **expanded, distended, dilated** or **inflated** consciousness or mind called **cosmic space;** god needs in order to amuse, entertain or regale itself whenever it feels lonely and unloved as is the case at present.

Fundamentally there are only **two things** in the current cosmos namely, consciousness or mind on one hand and **physical, objective** or **concrete matter** on the other. All the other things of the current cosmos have been formed out of these two fundamental ingredients of the current cosmos. There is no third thing in the current cosmos.

Consciousness or mind is eternally **dimensionless** whereas physical, objective or concrete matter is and will be

dimensional as long as it is allowed to exist by its creator namely, **cosmic space** aka god.

That is to say, consciousness or mind is eternally **non-dimensional** whereas physical, objective or concrete matter is and will be **3-D** or **three-dimensional** as long as it is permitted to exist by its **creator** namely god aka **cosmic space.**

Furthermore, consciousness or mind is always a 'non-created' thing whereas physical, objective or concrete matter is always a 'created' thing.

Consciousness or mind; irrespective of whether it is the eternally **bodiless** or **non-embodied** consciousness or mind of **cosmic space** aka god or that of an **embodied** consciousness or mind, for example, human consciousness or mind; is intrinsically and eternally endowed with the power to create physical, objective or concrete matter inside itself via the medium of its intrinsic and eternal ability of **daydreaming, oneiricking** or **oneirism.**

Physical, objective or concrete matter on the other hand, has no power to create anything. It can create neither physical, objective or concrete matter nor, consciousness or mind.

Insentient & **3-D** or **three-dimensional** physical, objective or concrete matter is merely a **play-thing** in the hands of god which the latter creates inside its own eternally **dimensionless** or **non-dimensional** consciousness or mind out of a **piece, portion** or **chunk** of its own consciousness or mind as and when it desires or fancies in order to amuse, entertain or regale itself i.e. when it feels lonely and unloved, nothing more nothing less.

Let one explain what one means when one says that the **insentient** and **3-D** or **three-dimensional** physical, objective or concrete matter is merely a **play-thing** in the hands of god which the latter creates inside its own eternally **dimensionless** or **non-dimensional** consciousness or mind out of a **piece, portion** or **chunk** of its own consciousness or mind as and when it desires or fancies in order to amuse, entertain or regale itself i.e. when it feels lonely and unloved, nothing more nothing less.

First and foremost, let one inform to oneself once again that the eternally **dimensionless** or **non-dimensional** plus **bodiless** or **non-embodied** consciousness or mind of god; in its **expanded, distended, dilated** or **inflated** form or version; is the entity which is ignorantly or delusionally called **cosmic space** by the **embodied** human consciousness.

This cosmic space aka the eternally dimensionless or non-dimensional plus bodiless or non-embodied consciousness or mind of god; in its expanded, distended, dilated or inflated form or version; has created a lump of insentient and 3-D or three-dimensional, physical, objective or concrete matter inside itself by condensing, compacting or congealing a portion, piece or chunk of itself through the activity of its daydreaming, oneiricking or oneirism.

This **insentient** and **3-D** or **three-dimensional** lump of **physical, objective** or **concrete** matter; thus created by this **cosmic space** aka god inside itself through its activity of **daydreaming, oneiricking** or **oneirism;** has been used by it to form, create or shape an extraordinary **array** or **assemblage** of **variety, diversity, heterogeneity** or **multiplicity** inside its **expanded, distended, dilated** or

inflated but unremarkable, uninteresting, nondescript or featureless consciousness or mind in order to amuse, entertain or regale itself i.e. when it feels lonely and unloved, nothing more nothing less.

Let one enumerate the other innate attributes of consciousness or mind, irrespective of whether it is the eternally **bodiless** or **non-embodied** consciousness or mind of **cosmic space** aka god or that of an **embodied** consciousness or mind, for example, human or animal consciousness or mind.

Consciousness or mind; irrespective of whether it is the eternally **bodiless** or **non-embodied** consciousness or mind of **cosmic space** aka god or that of an **embodied** consciousness or mind, for example, human or animal consciousness or mind; is **timeless, immortal** or **eternal.**

Insentient and **3-D** or **three-dimensional** physical, objective or concrete matter, on the other hand; irrespective of whether it has been used by god aka **cosmic space** to form absolutely **insentient** or incapable of feeling or understanding things, for example, **mountains, moons, planets, stars, galaxies** and the like or, it has been used to form the **physical, objective** or **concrete** bodies of **sentient** or capable of feeling or understanding beings, for example, human beings; is always **time-bound, mortal** or **non-eternal** and never **timeless, immortal** or **eternal.**

The duration of existence of **time-bound, mortal** or **non-eternal** plus **insentient** and **3-D** or **three-dimensional** physical, objective or concrete matter depends entirely upon the mood, whim or fancy of the **timeless, immortal** or **eternal** consciousness or mind namely **cosmic space** aka god who has created this **time-bound, mortal** or **non-eternal** plus **insentient** and **3-D** or **three-dimensional** physical, objective or concrete matter inside itself out of a **piece, portion** or **chunk** of its own consciousness or mind via the medium of its intrinsic and eternal ability of **daydreaming, oneiricking** or **oneirism.**

One must keep the following vital fact always in mind.

That the time-bound, mortal or non-eternal plus insentient and 3-D or three-dimensional physical, objective or concrete matter is merely a play-thing in the hands of timeless, immortal or eternal plus eternally dimensionless or non-dimensional consciousness or mind which the latter creates inside its own consciousness or mind as and when it fancies or desires in order to form, shape, produce or bring into existence an extraordinary array or assemblage of variety, diversity, heterogeneity or multiplicity inside its expanded, distended, dilated or inflated but unremarkable, uninteresting, nondescript or featureless consciousness or mind so as to amuse, entertain or regale itself when it feels lonely and unloved, nothing more nothing less.

Timeless, immortal or **eternal** plus eternally **dimensionless** or **non-dimensional** consciousness or mind; irrespective of whether it is the eternally **bodiless** or **non-embodied** consciousness or mind of **cosmic space** aka god or that of an **embodied** consciousness or mind, for example, **embodied** human consciousness or mind; creates inside itself, a quantum of **time-bound, mortal** or **non-eternal** plus **insentient** and **3-D** or **three-dimensional** physical, objective or concrete matter through the instrumentality of **daydreaming** or **oneiricking** on its part or, if it is preferred, through the instrumentality of **daydreamism** or **oneirism** on its part.

The ability to **daydream** or, if it is preferred, the ability to **oneiric** is a competence, capacity or power which is **innate** to all consciousnesses or minds and hence, it is possessed by all consciousnesses or minds.

That is to say, the ability to **daydream** or **oneiric** is an **innate** competence, capacity or power which is possessed by consciousness or mind of god aka **cosmic space** as well as consciousness or mind of man.

Timeless, immortal or eternal plus eternally dimensionless or non-dimensional consciousness or mind of god aka cosmic space creates inside itself a quantum of time-bound, mortal or non-eternal plus insentient and 3-D or three-dimensional physical, objective or concrete matter through the instrumentality of daydreaming or oneiricking on its part or, if it is preferred, through the instrumentality of daydreamism or oneirism on its part in order to form, produce or bring into existence an extraordinary array or assemblage of variety, diversity, heterogeneity or multiplicity inside its unremarkable, uninteresting, nondescript or featureless expanded, distended, dilated or inflated consciousness or mind aka cosmic space with the assistance or participation of the said quantum of time-bound, mortal or non-eternal plus insentient and 3-D or three-dimensional, physical, objective or concrete matter.

Since the **time-bound, mortal** or **non-eternal** plus **insentient** physical, objective or concrete matter is always a **dimensional** thing and never a **non-dimensional** or **dimensionless** thing or, better still, since the **time-bound, mortal** or **non-eternal** plus **insentient** physical, objective or concrete matter is always **3-D** or **three-dimensional** in shape, silhouette or contour and never **non-dimensional** or **dimensionless**; irrespective of whether it is in its **un-expanded, un-distended, un-dilated** or **un-inflated** form or version or, **expanded, distended, dilated** or **inflated** form or version; and since it cannot have an existence independent of its creator's consciousness or mind because it merely is a **daydream** or **oneiric** of its creator's consciousness or mind; it always, of necessity or perforce, occupies a certain amount of **consciousnessbal-space** or **mind-space** as per its **dimension, measurement** or **size,** inside its creator's **timeless, immortal** or **eternal** and **sentient** plus eternally **dimensionless** or **non-dimensional** consciousness or mind.

This consciousnessbal-space or mind-space; situated inside the timeless, immortal or eternal and sentient plus eternally dimensionless or non-dimensional consciousness or mind of the creator of the dimensional or, 3-D or three-dimensional physical matter, is given the name or epithet of cosmic space by the deluded, duped or hoodwinked embodied human consciousness.

Existence of **consciousnessbal-space** or **mind-space** inside the creator's **timeless, immortal** or **eternal** and **sentient** plus eternally **dimensionless** or **non-dimensional** consciousness or mind is a sine-qua-non or an indispensable pre-condition or pre-requisite for the birth and continued existence of the **time-bound, mortal** or **non-eternal** plus **insentient** and **3-D** or **three-dimensional,** physical, objective or concrete matter inside creator's consciousness or mind.

Time-bound, mortal or **non-eternal** plus **insentient** and **3-D** or **three-dimensional,** physical, objective or concrete matter is a **consciousnessbal-product** or **mind-product** only or, if it is preferred, is a **subjective-product** only, and never, under any condition, an **objective-product** as fallaciously or delusionally, believed and accepted by the **embodied** human consciousness.

The above statement is an **absolute truth.** Before offering the reason why the above statement is an **absolute truth,** it will be worthwhile for one to quote it once again which is as follows.

" The **time-bound, mortal** or **non-eternal** plus **insentient** and **3-D** or **three-dimensional,** physical, objective or concrete matter is a **consciousnessbal-product** or **mind-product** only or, if it is preferred, is a **subjective-product** only and never, under any condition, an **objective-product** as fallaciously or delusionally believed and accepted by the **embodied** human consciousness".

The reason why the above statement is an absolute truth is that the stark, blunt, unadorned, unembellished, or unvarnished fact is that the time-bound, mortal or non-eternal plus insentient and 3-D or three-dimensional, physical, objective or concrete matter has been created by the timeless, immortal or eternal plus eternally dimensionless or non-dimensional and non-physical, non-material, non-objective or non-concrete consciousness or mind of its creator aka god aka cosmic space inside itself through its activity of daydreaming, oneiricking or oneirism only and nothing else for the purpose of its amusement, entertainment or regalement only when it feels lonely and unloved, and nothing else.

COSMIC SPACE IS GOD AND NOTHING BUT GOD

On account of their delusion, many **embodied** human consciousnesses say:- 'God is nowhere'.

But truth is quite the opposite. God is all around them, above them, below them, by the side of them and also inside them. This god who, they say, is 'nowhere', is everywhere in the **physical** cosmos. There is no place in the **physical** cosmos where this god is not present. And to top it all, this god is not invisible. On the contrary, it is visible to all **embodied** human consciousnesses through the media of the eyes of their **physical** body.

Let one explain who this god of the **physical** cosmos is, god of the **physical** cosmos who, contrary to the popular belief amongst the **embodied** human consciousnesses, is truly visible and not invisible as surmised or conjectured by the **embodied** human consciousnesses for past countless millenniums.

This god of the **physical** cosmos is the entity whom **embodied** human consciousnesses call or give the epithet or moniker of **cosmic space** on account of their immense delusion.

The absolute truth is that **cosmic space** is god and nothing but god.

Since the **non-embodied, non-physical** and **immortal** cosmic space does not speak to **embodied** human consciousnesses in the **'human way'** or, if it is preferred, since the **non-embodied, non-physical** & **immortal** cosmic space does not communicate with the **embodied** human consciousnesses in the **'human way',** the **embodied** human consciousness must not be in the misapprehension, misunderstanding or miscalculation that this **non-embodied, non-physical** and **immortal** cosmic space is some kind of unexplainable, mysterious and **insentient** or incapable of feeling or understanding thing with which they cannot interact in any way, or, about which they cannot have any kind of vibes or, by which they cannot be affected in any way nor any **physical** object can react with it in any way or be affected by it in any way.

Cosmic space is perceivable and experienceable through the medium of only one sense organ of the **physical** body of the **embodied** human consciousnesses namely, the eyes of their **physical** body. Because of this **unique** characteristic of god of the **physical** cosmos aka **cosmic space, embodied** human consciousnesses are able to discern the existence of god of the **physical** cosmos aka **cosmic space.** Otherwise, **embodied** human consciousnesses cannot reach out to god of the **physical** cosmos aka **cosmic space** by any other physical or sensual means, for example, through their physical sense organs of touch, taste, smell or hearing.

Since the **non-embodied, non-physical** and **immortal** cosmic space aka god of the **physical** cosmos does not **communicate** or **interact** with the **embodied** human consciousnesses in the '**human-way**', **embodied** human consciousnesses have adopted the attitude that their best option in regards to dealing with this mysterious **cosmic space** should be the following :- Not to think about it and leave it alone to one side or to leave it to its own devices or fate and concentrate their energy plus focus on the solid, concrete, objective or physical matter of the cosmos only, the solid, concrete, objective or physical matter of the cosmos which is perceivable and experienceable to them through the media of all the five sense organs of their physical body and not through the medium of merely one sense organ namely the eyes as is the case with the mysterious non-embodied, non-physical and immortal cosmic space.

Furthermore, the **embodied** human consciousnesses find that the **solid, concrete, objective** or **physical matter**, even though it does not **communicate** with them in a **'human-way',** at least its saving grace or redeeming feature is that it can be put under the **light microscope** and **electron microscope** for physical or objective scrutiny or examination or, in a **test tube** for chemical or biochemical analysis or, in a **Large Hadron Collider** for smashing it to bits to the level of the smallest possible subatomic particles in order to solve the mystery of origin of the **physical matter**. No such treatment can be meted out to the **non-physical non-embodied, immortal** and immensely mysterious **cosmic space** by the **embodied** human consciousnesses. Therefore, the unwritten attitude of the **embodied** human consciousnesses towards

this **non-physical, non-embodied, immortal** and immensely mysterious **cosmic space** has been and still is :-
'To ignore it or to close one's eyes to it',
'Not to talk about it',
'Not to think about ',
'Not to waste 'valuable' human time on it', etc. etc.

The above described attitude of absolute neglect of **cosmic space** as a **serious subject** of study on the part of the **embodied** human consciousnesses is not a recent phenomenon. The same attitude has prevailed amongst the **embodied** human consciousness from time immemorial.

The above attitude of the **embodied** human consciousnesses towards the **non-physical, non-embodied, immortal** and immensely mysterious **cosmic space** has produced a very unfortunate consequence for many of them namely, that many of this class or category of **embodied** human consciousnesses, when asked **'Where is god?'**, they reply, **'God is nowhere'**, or **'There is no god'.**

They further add, that the **non-physical** and immensely mysterious human **consciousnesses** of the **physical** cosmos and, of course, the **non-physical** and immensely mysterious **cosmic space** of the **physical** cosmos, which is the subject of the present discussion, are both mere **by-products** or, mere **incidental** of **secondary products** or, mere **secretions** or **excretions,** which have emanated from or, have poured out of or, have exuded out of the absolutely **insentient** or the absolutely incapable of feeling or understanding **physical matter** of the **objective** or **physical** cosmos and nothing else, because in their view or opinion or, better still, as per their theory, the absolutely **insentient** or the absolutely incapable of feeling or understanding **physical**

matter, even though without doubt is a one hundred percent **insentient** or one hundred percent incapable of feeling or understanding thing, is nevertheless the **supreme being,** the **ultimate being,** the **highest ranking being** or, the **absolute being** or, better still, is the **absolute monarch** of the **physical** cosmos as well as the **absolute monarch** plus the **source** or the **fountainhead** of the **non-physical** and immensely mysterious human **consciousnesses** of the **physical** cosmos on one hand and, of course, the **source** or the **fountainhead** of the **non-physical** and immensely mysterious **cosmic space** of the **physical** cosmos on the other. The **non-physical** and immensely mysterious **cosmic space** of the **physical** cosmos on one hand and the **non-physical** and immensely mysterious plus **sentient** or capable of feeling, understanding, thinking, analyzing, questioning and daydreaming or oneiricking human **consciousnesses** of the **physical** cosmos on the other, are both subordinate or subservient to the absolutely **insentient** or absolutely incapable of feeling or understanding **physical matter** of the **objective, concrete** or **physical** cosmos.

They further imply, without saying so in so many words that the **non-physical** and immensely mysterious **cosmic space** on one hand and the **non-physical** and equally mysterious plus **sentient** or capable of feeling, understanding, thinking, analyzing, questioning and daydreaming or oneiricking human **consciousnesses** of the **physical** cosmos on the other, have both been **excreted out, secreted out, exuded out** or **poured out** by the absolutely **insentient** or absolutely incapable of feeling or understanding **physical matter** of the **objective** or **concrete** cosmos in order **to merely play second fiddle** or **subsidiary role** to the absolutely insentient or absolutely incapable of feeling or understanding **physical matter** of the **objective** or **concrete** cosmos.

What has been said above can be put in another way.

As per those **embodied** human consciousnesses who are the staunch or steadfast 'nay-sayers' or 'anti' with regards to the existence of god, the absolutely insentient or absolutely incapable of feeling or understanding **physical matter** of the **objective** or **concrete** cosmos is the **source, fountainhead** or **mother** of everything in the **physical** cosmos which includes the **non-physical** and immensely mysterious plus **sentient** or capable of feeling, understanding, thinking, analyzing, questioning and daydreaming or oneiricking human **consciousnesses** on one hand and equally mysterious plus non-physical **cosmic space** on the other.

However, the stark or the blunt truth is that without the pre-presence or pre-existence of immensely mysterious and non-physical **cosmic space,** the subsequent **birth** and **spatial-placement** some 13.7 billion light years ago plus the continued **spatial-existence** of the **3-D** or **three-dimensional** physical matter of the **objective** or **concrete** cosmos would have never materialized. This stark or blunt fact has been totally overlooked by those **embodied** human consciousnesses who are the staunch or steadfast 'nay-sayers' or 'anti' with regards to the existence of god and who fervently bat for the **theory-of-supremacy** of the absolutely insentient or absolutely incapable of feeling or understanding **physical matter** in the formation of all the **player** who or which are currently taking part in the countless affairs of the **objective** or **concrete** cosmos in order to make it memorable.

The gross, portly or bulky plus absolutely insentient or absolutely incapable of feeling or understanding **physical matter** of the objective, concrete or physical cosmos can never be the **source, fountainhead** or **mother** of the mysterious **cosmic space** nor can it ever be the **source, fountainhead** or **mother** of the equally mysterious human and animal consciousnesses of the **physical** cosmos, the mysterious human and animal consciousnesses both of whom are **sentient** beings in **stark** or **blunt** contrast to the absolutely **insentient** physical matter of the objective, concrete or physical cosmos.

It is the mysterious **non-embodied, non-physical** and **immortal cosmic space** and **cosmic space** alone who can be and in fact is the **source, fountainhead** or **mother** of the gross, portly or bulky plus absolutely insentient or incapable of feeling or understanding **physical matter** of the objective, concrete or physical cosmos and not the other way around as theorized by those **embodied** human consciousnesses who are the staunch or steadfast 'nay-sayers' or 'anti' with regards to the existence of god and are fervent backers of the **theory-of-supremacy** of the gross, portly or bulky plus absolutely insentient or incapable of feeling or understanding **physical matter** in the **objective** or **concrete** cosmos.

Even those **embodied** human consciousnesses who think that there is god, they say and **deeply** believe that this god lives far away from its creation in some mysterious and exalted place called **heaven** or whatever. What an extremely unfortunate situation?

The tragedy which has befallen or, has come upon the analytical section, sector, area, region or zone of those **embodied** human consciousness who **deeply** believe that god lives far away from its creation in some mysterious and exalted place called **heaven** or whatever is that they visualize or conceive god in terms of someone whose dimensions or aspects are **physical, material, substantial, objective** or **concrete.** Otherwise, they would have not thought or conceived that god lives in some far away and exalted **place** called **'heaven'** or whatever, meaning thereby, otherwise, they would have never thought or conceived that god lives separate from its creation.

What has been said above can be put in another way.

Terms like **'heaven'** or whatever, chosen by these **embodied** human consciousnesses as being the **abode of god** or **home of god** creates confusion in the **analytical** section, sector, area, region or zone of **embodied** human consciousness in the sense that they believe or conceive that some dimensions or aspects of god are related to **physicality.** This confusion in the **analytical** section, sector, area, region or zone of these **embodied** human consciousnesses has led them astray or led them away from the correct path with regards to their search for the answer to such questions as :-

1. **Who** is the **creator** of the **physical** cosmos?
2. **How** the **creator** of the **physical** cosmos brought into being the **physical** cosmos?
3. **Why** the **creator** of the **physical** cosmos brought into being the **physical** cosmos**?**

The above questions are not the only questions which confront **embodied** human consciousnesses vis-a-vis the **physical** cosmos. There are countless other doubts or uncertainties in the **analytical** and **questioning** section of the **embodied** human consciousnesses which need to be answered. These will be tackled in a later volume.

~*~*~*~*~

ABOUT THE AUTHOR

Dr. Chandra Bhan Gupta, was born and educated in Lucknow, India.

He commenced his medical career in India with several notable medical articles to his credit.

Subsequently, he went to UK., where he continued his distinguished medical career, gaining the highest postgraduate and honorary accolades within his field.

Such questions as how and why man and the rest of creation have come into being, as well as the true nature of the creator and where is his abode, troubled him from an early age.

In an attempt to find answers to these eternal questions, he went through extreme austerities or penance over the course of many years, accompanied by long periods of deep meditation.

Enlightenment from the Almighty came in 1995, which resulted in the writing of the first book on the theme of "Supra-Spirituality", called **'Adwaita Rahasya: Secrets of Creation Revealed',** followed by two more books delving deeper into the same theme, entitled **'Space is The Mind of God: A Scientific Explanation of God and His Abode',**
and more recently the first of a series of works, **entitled 'Cosmic Space is God and Universe is God's Dream'.**

Printed in Poland
by Amazon Fulfillment
Poland Sp. z o.o., Wrocław

54779740R00148